Puzzlers' Tribute

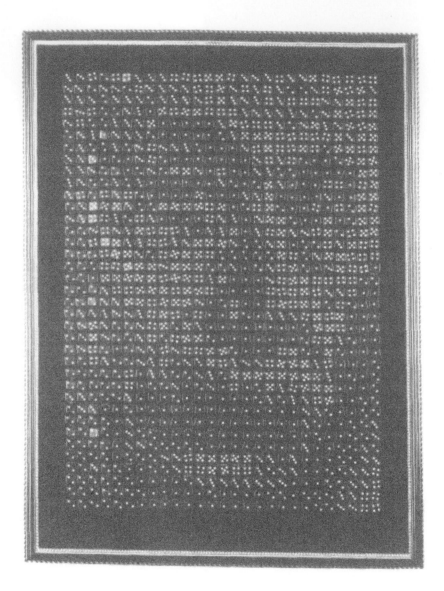

"I shall never believe that God plays dice with the world." — Albert Einstein

Puzzlers' Tribute
A Feast for the Mind

Edited by

David Wolfe and Tom Rodgers

CRC Press
Taylor & Francis Group
Boca Raton London New York

CRC Press is an imprint of the
Taylor & Francis Group, an **informa** business

AN A K PETERS BOOK

CRC Press
Taylor & Francis Group
6000 Broken Sound Parkway NW, Suite 300
Boca Raton, FL 33487-2742

© 2002 by Taylor & Francis Group, LLC
CRC Press is an imprint of Taylor & Francis Group, an Informa business

First issued in paperback 2019

No claim to original U.S. Government works

ISBN 13: 978-0-367-44715-1 (pbk)
ISBN 13: 978-1-56881-121-5 (hbk)

Visit the Taylor & Francis Web site at
http://www.taylorandfrancis.com

and the CRC Press Web site at
http://www.crcpress.com

Contents

Preface

Mathematicians, magicians, and puzzlists are masters of the unsolvable, the unbelievable and the undoable. Their currency is paradox. They reach enlightenment through bewilderment.

Members of these three communities meet every two or three years to honor the man at the forefront and nexus of all three, Martin Gardner. At this remarkable Gathering for Gardner (the four so far are dubbed G4G1 through G4G4), participants share talks and performances; problems and puzzles; knowledge and ideas. We invite you to read this compendium of contributions so that you too can be bewildered and enlightened.

The Mathemagician and Pied Puzzler was published in 1999 as a tribute to Martin Gardner based on contributions to G4G1. Many of the Gatherings' participants and other fans of Martin Gardner expressed the wish for a second volume. We were glad to have another opportunity to share some of group's favorite paradoxes, problems and puzzles. The articles enthusiastically proffered by G4G participants honoring Gardner have led to the creation of a website and the start of a third volume. Emily DeWitt Rodgers set up and maintains the dedicated website, http://www.g4g4.com, which includes the full text of the first tribute book and a maze of puzzles, illusions, and problems. Many contributions and materials are placed there with the permission of the authors along with links to homepages of G4G participants.

All of the things that we said in the introduction to *The Mathemagician and Pied Puzzler* remain true and relevant for this book so we have included words from its preface here:

Martin Gardner has had no formal education in mathematics, but he has had an enormous influence on the subject. His writings exhibit an extraordinary ability to convey the essence of many mathematically sophisticated topics to a very wide audience. In the words first uttered by the mathematician John Conway, Gardner has brought "more mathematics, to more millions, than anyone else." It is a moving testimony that many professional mathematicians feel that Martin Gardner sparked and guided their early interest in mathematics.

In January 1957, Martin Gardner began writing a monthly column called "Mathematical Games" in *Scientific American*. He soon became the influential center of a large network of research mathematicians with whom he corresponded frequently. On browsing through Gardner's old columns, one is struck by the large number of now-prominent names that appear therein. Some of these people wrote Gardner to suggest topics for future articles; others wrote to suggest novel twists on his previous articles. Gardner personally answered all of their correspondence.

Gardner's interests extend well beyond the traditional realm of mathematics. His writings have featured mechanical puzzles as well as mathematical ones, Lewis Carroll, and Sherlock Holmes. He has had a life-long interest in magic, including tricks based on mathematics, on sleight of hand, and on ingenious props. He has played an important role in exposing charlatans who have tried to use their skills not for entertainment but to assert supernatural claims. Although he nominally retired as a regular columnist at *Scientific American* in 1982, Gardner's prolific output has continued.

Martin Gardner's influence has been so broad that a large percentage of his fans had only infrequent contacts with each other, until Tom Rodgers conceived of the idea of hosting a weekend gathering in honor of Gardner to bring some of these people together. The first "Gathering for Gardner" (G4G1) was held in January 1993. Elwyn Berlekamp helped publicize the idea to mathematicians. Mark Setteducati took the lead in reaching the magicians. Tom Rodgers contacted the puzzle community. Out of this first gathering grew a serious of events; a second gathering, G4G2, was held in January 1994, G4G3 in January 1998, and G4G4 in February 2000.

The success of these gatherings has depended on the generous donations of time and talents of many people. The organizers, Elwyn Berlekamp, Tom Rodgers, Mark Setteducati would like to acknowledge the work of many people who have helped make the Gatherings for Gardner successful, including Scott Kim, Jeremiah Farrell, Karen Farrell, Emily DeWitt Rodgers, David Singmaster, and many others.

Of course, this book could not exist without the efforts of a large group of contributors. Scott Kim conceived of and assembled the first tribute volume, and created the graphically spectacular cover in this second volume; Scott's generous contribution will remain engraved in our memories. Emily DeWitt Rodgers has done an excellent job of designing and maintaining the g4g4.com website. In addition, we owe thanks as well to a number of anonymous reviewers for reviewing articles outside the realms of expertise of the editors. David Wolfe is indebted to his wife, Susan Hirshberg, for her incredible support and writing expertise, even while he negected wedding and honeymoon plans to work on this book. As with any book, only the competence and professionalism of our publisher, A K Peters, Ltd., has allowed this project to overcome the difficult transition from an idea spoken of over wine to a real book which is now in your hands.

All of us feel honored by this opportunity to join together in tribute to the man in whose name we gathered, Martin Gardner.

David Wolfe   Tom Rodgers
St. Peter, Minnesota   Atlanta, Georgia

David Klarner

Harry Eng

PHONE (204) 555-1234

FAX (204) 555-5678

melStover

Mel Stover

A Playful Tribute—Scott Kim

In Memoriam

Even mathemagicians are mortal. This book is dedicated to the memory of three participants in The Gathering for Gardner, Mel Stover, Harry Eng, and David Klarner. Each was an admirer of Gardner and has gathered into the folds of another dimension since last we met.

Mel Stover, a fifty-plus-year friend and correspondent of Martin Gardner, lived magic and created his own deliciously pernicious magic-puzzles. Both Martin Gardner and Max Maven wrote articles not only honoring Mel, but also remembering many of Mel's illusions.

"The Impossible Just Takes Longer" was oft quoted by Harry Eng, whose magic tricks, impossible objects, and memory feats pushed the limits of man's capabilities, as described in Mark Setteducati's article. The solution of Harry Eng's Impossible Bottle with inserted coins larger than the bottle mouth is explained in Gary Foshee's article.

David Klarner, who is remembered in articles by Solomon Golomb, C. J. Bouwkamp, and David Singmaster, helped develop the mathematics of box-packing problems and created masterful box-packaging puzzles.

We will miss them.

The Participants

G4G1, G4G2, G4G3, G4G4

A Greeting to Martin Gardner

Sir Arthur Clarke CBE

I am very happy to send my greetings to Martin on the occasion of the 'gathering' in his honour.

In particular, I notice that the theme of the event is the "Fourth Dimension"—something that has always fascinated me. I had almost forgotten that is was featured in one of my earliest (1946!) stories *Technical Error* (now in *Reach for Tomorrow*). And my very first television programme (BBC TV, 4 May 1950) was a thirty minutes talk on the Fourth Dimension—live of course, because there was no video tape in those days! After that ordeal, no camera has ever had any fears for me.

I am very grateful to Martin for creatively disrupting my life on at least two occasions, thanks to his columns in *Scientific American*. Thirty years ago he turned me on to Pentominos, with results you'll see in *Imperial Earth*. However, even more important, he opened my eyes to the infinite universe of the Mandelbrot Set, which I cleverly managed to combine with SS Titanic, in *The Ghost from the Grand Banks*. (Isn't the connection obvious?)

Finally, it was Martin's *The Night is Large* that inspired me to put together my own collection of non-fiction, *Greetings, Carbon-Based Bipeds!*

I would be hard put to think of anyone else to whom I owe so great an intellectual debt, and I wish him many more years of happy puzzling!

6 December 1999
Colombo, Sri Lanka

Sir Arthur Clarke

Part I
The Toast Tributes

Harry Eng: A Tribute

Mark Setteducati

Harry Eng was a teacher, inventor, minister, artist, magician, and musician whose life was about thinking and inspiring others to think. He was born in California, but the details are vague because he was adopted. For more than 30 years he lived in La Mesa with his wife Betty, raising two children, Greg and Diana.

Harry's house is filled with his creations and described by all who visit as the most unique home they've been in. An extraordinarily innovative school teacher in the San Diego area, in recent years he acted as a consultant to schools giving special lectures to gifted students on creativity and thinking. Harry created many original teaching techniques and devices such as using play money in class printed with his students' pictures on it, and a dummy named Ziggy that had clear plastic lungs attached to a vacuum cleaner. Harry would light up a cigarette for Ziggy and turn on the vacuum to teach kids not to smoke. He used magic and more recently his impossible bottles to inspire students to think.

Harry had a lifelong interest in magic. During the late 1970s he was actively involved in the San Diego magic scene, including being president of a local IBM ring. He never read books on magic and everything Harry performed was original. He invented the "PK Factor," a magnetic principle that was marketed by magic dealers. His hands were chubby and twisted, yet out of these would come incredible feats of magic, inventions, and impossible bottles. Magicians would show him a trick and Harry would often have an insight on a better way to do it—always in a humble way. Harry never made anybody feel he was better or smarter.

Mark Setteducati is a magician and inventor of magic, games and puzzles. He created Milton Bradley's *Magic Works* ® and is co-author of *The Magic Show*. This article originally appeared in *Genii* in October, 1996.

One of Harry's passions was his feats of memory which included books he made that contained 10,000 numbers or thousands of words that he had memorized. He would have you turn to any page and he would recite the contents without looking. He created a cardboard name computer that has over 1000 names programmed into a few cards. With a few questions he could guess anybody's name. He created his own stacked deck and was constantly working on new and more impossible card effects with it. He was a master at origami and would fold an ordinary paper bag into a hat, a pair of shoes or a wallet. My favorite routine that Harry performed was with a single piece of heavy duty rope, sometimes with his trademark button knot tied at the bottom. He would pass the rope behind his back while talking about a courtroom. The first time a knot would appear in the center of the rope. Harry would say, "Knot Guilty." The second time Harry would amazingly be able to snap the rope behind his back so it created a noose as he would exclaim, "Guilty!... Hang 'em!" He would then go on to thread a needle, shoot the rope as a bow and arrow, penetrate the rope through a spectator's thumbs, and arrange the rope between his fingers to create a frog and then a dog. Each trick would be accompanied by silly patter and clever puns, all executed with precision skill and a sense of humor that exemplified Harry's personality.

Figure 1.   Harry Eng's bottles. Harry explained that cutting the bottle is the hard way to fit the items in for it's nearly impossible to disguise the cut when resealing. (See Color Plate II.)

And then there are the bottles. If Harry never made a single bottle he would still be one of the most remarkable men I have known—but it's the bottles that certify him as a true genius. Impossible bottles date back more than a hundred years, and Hoffman describes an arrow through a bottle, as well as a bottle containing a dowel with a screw through it. And of course there is the classic ship in a bottle. But about ten years ago Harry Eng put a deck of cards in a bottle and it went on from there: ping pong balls, tennis balls, coins, sneakers, padlocks, baseballs, light bulbs, scissors.... You name it and Harry has probably put it in a bottle. He brought to his bottles a level of originality and diversity that astonished and delighted top puzzle experts in the world. The beauty of the way he would tie knots inside the bottles using the same rope he would perform his tricks with and the clever puns he would incorporate in many of the bottles (a deck of cards with a bullet through it is titled a "loaded deck") make them more than just puzzles—they are works of art. When people look at the bottles Harry would say they became Indians because they always say "How." After giving one of his standard humorous answers such as "trained cockroaches" he would tell you his real secret which is: He "thinks his way into the bottle."

Most people assume the bottles were cut, which, according to Harry, would be the hard way to do it—there would be burn marks or evidence on the bottles. Then, after a little thought, people would figure he put the cards in one at a time or he took the lock apart and assembled it inside the bottle and would be satisfied with their answer. But it's exactly at this point, if you keep thinking about it, the more impossible they become. How did he get the steel nut through the deck—there's no clearance to screw the bolt on. How did he lock the padlock through the wooden plug capping the bottle—it doesn't move. Harry put things in bottles to challenge himself and to make people think.

After suffering a major heart attack more than 15 years ago, the doctors gave Harry about a year to live. Thankfully he didn't listen to them then—but he did suffer health problems that made him face his mortality every day, never letting it get in his way of living and enjoying life to the fullest. Ironically it was in the middle of performing his rope routine for a group of friends in Northern California on the afternoon of July 29, 1996 at age 64, Harry felt a little faint, sat down and passed away.

Harry truly loved people. He traveled extensively and had friends all over the world—there isn't a person who knew him who hasn't been influenced by his inspirational mind and personality. He was one of those rare people who I never saw get angry and who would always look at the positive side of things. Since his passing I think of him every day and the image that keeps coming to me is not of his incredible mind or creations, but of a happy man with a great sense of humor that was reflected in everything he did—always telling silly jokes and laughing. Harry loved to laugh.

The Eng Coin Vise

Gary Foshee

Of all the objects Harry Eng has placed inside of a bottle, a solid metal coin is one of the most amazing. The effect is one of total impossibility. I pondered over this for some time until Harry let me in on his secret: he bent the coin, placed it in the bottle, then used a specially constructed metal vise *inside* the bottle to flatten the coin. To accomplish this, some of the vise pieces are fed into the bottle one at a time, and then reassembled while inside the bottle. The remaining pieces of the vise are assembled outside of the bottle and joined to the pieces inside. The bent coin is now maneuvered into the vise inside the bottle, the vise closed, and the coin pressed flat. The vise is now disassembled and removed from the bottle. The real beauty of the vise is that the force applied to flatten the coin is generated *outside* of the bottle.

The basic principles of Harry's vise are presented here. There is sufficient detail to build a vise from this description, but it will prove quite difficult. Proper high-strength steels must be used for certain components to withstand the considerable force required to flatten the coin. Consulting someone skilled at metalworking is highly recommended.

The vise consists of an inside piece, an outside piece, and a connecting rod joining the inside and outside pieces. The inside piece is shown unassembled in Figure 1 and assembled in Figure 2. The central hole in the upper bar of Figure 1 is threaded to accept the connecting rod. The two lower bars of Figure 2 slide freely on the two bolts.

The parts of the outside piece are shown in Figure 4. The block has a vertical hole through the center. The hole is threaded at the top to accept a bolt, and threaded at the bottom to accept the connecting rod. A solid push rod fits inside of the connecting rod. The part to the right

---

Gary Foshee is a mechanical puzzle collector and designer living in Seattle, Washington.

Figure 1.

Figure 2.

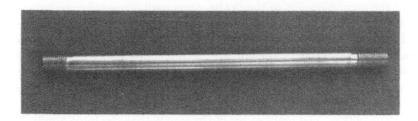

Figure 3.

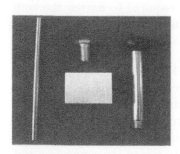

Figure 4.

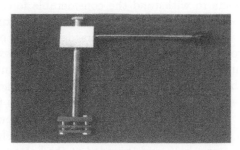

Figure 5.

Figure 6. Figure 7.

of the block is a handle that threads into a hole in one side of the block. The connecting rod is shown in Figure 3.

To assemble the vise, feed the parts of the inside piece into the bottle, and join them as shown in Figure 2. This will require considerable dexterity, patience, and many hours of practice. Join the inside and outside pieces with the connecting rod. Next, slip the push rod in through the top of the block and push it all the way down into the inside piece. Thread the bolt into the upper part of the block. Attach the handle. The vise should now appear as in Figure 5.

Place the bent coin in the bottle and maneuver it into the vise as shown in Figure 6. This is very difficult because the connecting rod keeps getting in the way. Place a wrench on the bolt and tighten, gripping the handle with your other hand. This force will be transmitted via the push rod to the coin, and flatten it as shown in Figure 7. Several pressings are needed to completely flatten the coin. When the pressing is complete, disassemble and remove the vise, and your coin bottle is ready to amaze all that view it.

Figure 1

Figure 2

David A. Klarner—A Memorial Tribute

Solomon W. Golomb

I may have been the Prophet of Polyominoes, but David Klarner was their most faithful Apostle.

It was Christmastime in 1959 (or thereabouts) when I received a large oblong rectangular wooden box in the mail. The return address was one D. A. Klarner from the far north of California. The box had a sliding lid, which I opened carefully, and dumped the wooden contents out on a tabletop. These turned out to be the 29 "pentacubes," and it took several hours to get them back in the box as neatly as they had arrived. David's enclosed letter revealed that he was a student at Humboldt State College who had learned about polyominoes from Martin Gardner's columns in *Scientific American*.

Several years later, when I had gone from JPL to USC, I was invited by Professor Leo Moser of the University of Alberta, in Edmonton, Canada, to be the "outside reader" of the Ph.D. dissertation of one of his students. The student was David Klarner, and the doctoral thesis contained a proof that the number $P(n)$ of n-ominoes lies between a^n and b^n, where $a = 2$ and $b = 8$ was an acceptable choice. It was a day in mid-March in 1964 (or thereabouts) when I flew from Los Angeles around noon (80°F), and with several intermediate stops arrived in Edmonton late at night (-40°F) and met David for the first time. It had been a severe winter, and he had been ill with pneumonia much of the time. The next day was his successful thesis defense.

David had an important article, "Packing a Rectangle with Congruent N-ominoes," published in Volume 7 of the *Journal of Combinatorial Theory*, in 1969, where he introduced the concept of the **order** of a

---

Solomon W. Golomb is USC's University Professor, renowned for his work in shift register sequences, radar, number theory, fluency in numerous languages, and many other things.

11

polyomino, defined as the minimum number of congruent copies which can be assembled to form a rectangle. (If the given polyomino does not tile any rectangle, its order is undefined.) In the same article, he defined the **odd-order** of a polyomino to be the smallest odd number (if any) of congruent copies which can be assembled to form a rectangle. He had several beautiful illustrative examples, and the subject has inspired important research ever since.

When David was on the faculty at Stanford, he invited me to give a seminar talk (on polyominoes, of course). The date of my talk should be easy to establish, because the news item of the day was the death of former president Lyndon B. Johnson.

After Stanford, David was at SUNY-Binghamton, and also spent more than one sabbatical year visiting the Technical University of Eindhoven, in the Netherlands, where he interacted with N. G. de Bruijn and L. E. J. Bouwkamp, among others. From Binghamton, his next academic position was at the University of Nebraska, in Lincoln.

David had suffered since childhood from "type 1" diabetes (formerly called "juvenile onset" diabetes), and had a lifelong battle with the many complications of this ailment. Driving from New York to Nebraska, around the time he arrived in Lincoln he had a near-fatal heart attack. He was also the recipient of a kidney transplant.

It was in 1993 (or thereabouts) that I was invited (no doubt at David's suggestion) to be a member of a team of visiting experts to evaluate the progress of the University of Nebraska in the several areas of science and technology that the State Legislature had identified for special funding. I spent most of the free time I had during that period visiting with David Klarner—as it turned out, for the last time.

Ever since the first edition of *Polyominoes* appeared in 1965 (published by Charles Scribner and Sons), I was compiling material for a second edition, which finally appeared in 1994 (with Princeton University Press). I had considerable help from David in the preparation of the final manuscript for the new edition.

David had been the editor of the *Mathematical Gardner,* to which I contributed a chapter; and he was asked to edit the proceedings of the first "Gathering for Gardner", but by that time his health problems seriously interfered.

In mid-March of 1999, I was visiting the University of Waterloo, in Ontario, Canada, for several days. Douglas Stinson had been at the University of Nebraska when I had visited there, but was now at Waterloo. I asked him what he knew of David's whereabouts. He told me that David had retired, for health reasons, and had relocated, with his wife, back to Humboldt, California. I resolved to contact David when I returned to Los Angeles, but I never got the chance. Shortly after coming home, I learned the sad news of his demise.

David Klarner made an important and very distinctive contribution to the literature of combinatorial mathematics in general, and to polyominoes in particular. He will be long remembered by many mathematicians who never actually met him for the quality of his work, and he will be sorely missed by those of us who knew him.

David Klarner's Pentacube Towers

C. J. Bouwkamp

Dedicated to the memory of David A. Klarner

David, the Inventor: He writes, in a letter to Solomon W. Golomb dated November 27, 1959, "The box which this letter is attached to contains the mathematical blocks that I invented and named pentacubes. They are so named because they are all of the combinations of five cubes joined face to face." The total number of distinct pentacubes is known to be 29. The 12 planar pentacubes are also called solid pentominoes.

David, the Constructor: One of his many constructions with solid pentominoes is, what I call, the pentacube tower. To my knowledge, David never published or mentioned it to his friends before 1968. Among his many drawings, I was fortunate enough to find two different towers which are copied in Figure 1 (Type I). The reader, if in possession of a set of pentacubes, will certainly feel rewarded to successfully construct either tower.[1]

At the end of 1968, David came to Eindhoven University of Technology for the first time; his third and last visit to Eindhoven University was in 1991. He found colleagues and friends not only in Mathematics but also in the Building and Architecture Departments, and lectured about combinatorial problems.

C. J. Bouwkamp, from the Netherlands, is a long time reviewer for *Mathematical Reviews*. He has had lectureships at institutions throughout the U.S. and Europe.

[1]Pentomino aficionados assign each block a letter name. David used the letters C and S instead of Golomb's U and Z, probably because U might be taken for V, and similarly quarter-turned Z for N.

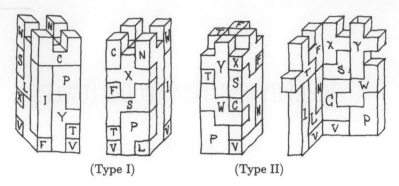

(Type I)   (Type II)

Figure 1.   On the left are two distinct pentacube towers of Type I due to Klarner.   On the right, Klarner's first known pentacube tower of Type II along with the same tower shown unfolded.

In September 1968, I became aware of the existence of yet another pentacube tower, as depicted in Figure 1 (Type II). Type I and Type II differ in the positions of the turrets, and both have an empty column at the inside. The reconstruction of the Type II tower is not that easy, so David added the fold-up diagram.

Once David was able to find one or two solutions to his own challenges, his interest quickly turned to other problems; finding all solutions was left to the computer. Type I has 10 solutions and Type II has 27, modulo reflection, as obtained by me on a "big" computer in October 1968. They are presented in plane diagrams in Figures 2 and 3, in terms of pentominoes.

The ⊏▢⁺⌐ indicates a ⌐⌐ turned on its side with the vertical stem pointing into the page. Similarly ⊏•—•⌐ is a ⌐⌐ turned on its side. How to convert the diagrams to towers is a puzzle in itself! The method is shown in Figure 4.

Figure 2.   The 10 pentacube towers of Type I.

Figure 3.   The complete set of 27 pentacube towers of Type II.

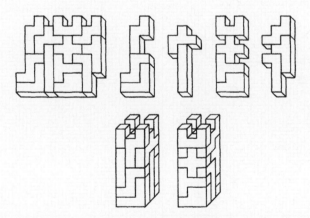

Figure 4.   The upper left is the first diagram of Type I from Figure 2. The diagram is split into four parts, folded, and reassembled to form the pentacube tower. Two views of the same tower are shown here.

Y2K Tribute to Martin Gardner

I'd like to close with a creation of my own.

Figure 5 shows a tiling of a square with 2000 congruent Y-pentominoes. The "Y" of Y2K refers both to the year and to the Y-pentomino. The tiling is simple (not compound), in that no subset of Y-pentominoes tiles a smaller rectangle. Moreover, it is rotationally symmetric, and does not contain any of the patterns shown below, patterns which are commonly used to aid in tilings:

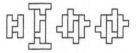

Rotations and reflections of these patterns are also forbidden.

The reader is invited to find the Y-pentomino patterns hidden in the tiling!

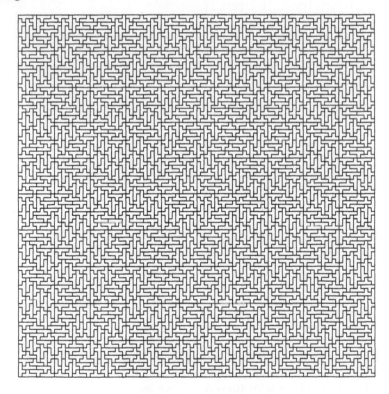

Figure 5.   Y2K tiling.

Some Reminiscences of David Klarner

David Singmaster

I first met David in 1970 or 1971, when he was visiting Reading University. We discussed box-packing and I found I had solved a problem that he'd been considering. In two dimensions, if a brick packs a box (larger than itself), then one can divide the box into two smaller boxes such that each smaller box can be packed (indeed with its bricks all in the same direction). The simplest illustration is filling a 6×5 box with 3×2 bricks, where the only packings exhibit this divisibility property.

David had wondered if this still held in three dimensions and I had found that 25 $1 \times 3 \times 4$ bricks can pack into a $5 \times 5 \times 12$ box but could not pack $5 \times 5 \times c$, for $c = 1, ..., 11$, nor $1 \times 5 \times 12$ nor $2 \times 5 \times 12$ in any way. This is the smallest example of this behavior. David later mentioned this in his classic "Brick-packing puzzles"[Kla73], but he cited an different example: $2 \times 3 \times 7$ in $8 \times 11 \times 21$, apparently having forgotten the numbers in my example.

In 1978, Dean Hoffman proposed the following. Can one fit 27 bricks, all $a \times b \times c$, into a cube of side $a + b + c$? The planar version is to use 4 bricks of size $a \times b$ to fit into a square of side $a + b$. This is easy to do and is a way of showing the arithmetic-geometric mean inequality

$$\sqrt{ab} \leq (a + b)/2$$

in the form

$$4ab \leq (a + b)^2.$$

The corresponding inequality for three variables gives us $27abc \leq (a + b + c)^3$ so that a solution of Hoffman's problem gives a geometric proof of the arithmetic-geometric mean inequality for three variables.

David Singmaster was the leading expositor of the mathematics of the Rubik's Cube and is presently the principal historian of recreational mathematics.

Hoffman tried to do this using pencil and paper and found it too hard to do, so he rang up David and asked if he could do it. David had inherited a fine table saw from his father and used it to make up a set of 27 blocks. As he made each one, he stacked it in the corner and found a solution as he went.

In the early 1980s, I visited David at Binghamton. Dean Hoffman was present and David made me a set of 27 blocks from a lovely redwood. He also made a three-cornered frame to hold them. This set is one of the treasures of my collection.

It was on this visit that I saw the bedspread made by Kara Lynn Klarner showing two orthogonal Latin squares of order 10. This is basically a 10 × 10 array of squares, using ten colors such that each color occurs once in each row and column. Then each square has a circle on it, using the same ten colors so that each color occurs once in each row and column, and further, so that each pair of colors occurs just once as a square-circle pair. They said the hardest part of making the spread was finding ten sufficiently contrasting colors.

Bibliography

[Kla73] David Klarner. Brick-packing puzzles. *Journal of Recreational Mathematics*, 6(2):112–117, 1973.

Just for the Mel of It

Max Maven

Let's get this part out of the way: In late March, 1999, he became the late Mel Stover. I'm not happy about that, and I'd guess that Mel wasn't particularly thrilled about it either.

When Houdini was hospitalized, he held out past the doctors' predictions and managed to die on Halloween. If Mel had held out just a few days longer, he'd have died on April Fool's Day. Then again, when Houdini was Mel's age he'd been dead for 34 years.

My introduction to Mel Stover was via Martin Gardner's seminal *Mathematics Magic and Mystery,* which I encountered a few years after its 1956 publication. There were several exceedingly clever contributions from Mel, including some of his groundbreaking work on geometric vanishes. I was quite taken with his trick entitled "Gargantua's Ten-Pile Problem," which required a deck of 10 billion cards. Mel's suggestion as to how such a pack could be most easily assembled: "Buy 200 million decks of 52 cards each, then discard two cards from each deck."

Clearly, this was no ordinary inventor.

Over the ensuing years, I came across Mel Stover's name in the pages of such learned journals as *Scientific American, The New Phoenix,* and *Ibidem.* And in the mid-1970s, I became friends with the Winnipeg wag himself.

As indicated above, Mel was a prolific contributor to a range of periodicals in the fields of puzzles, gaming, and magic. Most recently, during the last three years of the Larsen era, readers of this magazine were treated to his monthly "Braintwisters" column. In books, not

---

Orson Welles wrote that **Max Maven** has "the most original mind in magic." Fortuitously, Mr. Welles died before he could revise his opinion. This article first appeared in *Genii* in September, 1999. The text is copyright by Max Maven, and used with permission.

infrequently, his contributions were simply appropriated. That was all the more unfortunate, given that Mel was a particularly generous fellow who, in most cases, surely would have granted permission had it been sought.

Despite this general largesse, there were certain creations Mel held back from releasing to a wider audience, preferring to retain the option of using them to bedevil friends and acquaintances.

Mel was especially fond of devious variants on old puzzles. The scenario would usually run something like this:

MEL: "Say, y'ever seen this one?"

VICTIM: "Um, yeah, I used to know that, years ago. Let me see, I think I sort of vaguely remember..."

And that was it. You were screwed. Because whatever you vaguely remembered, what Mel had given you was different. Oh, it looked the same—at least, you thought it did—but it wasn't.

Of these mischievous pranks, Mel's proudest accomplishment was clearly his take on the classic horse-and-rider puzzle. The origin is a type of ambiguous novelty picture that dates back at least as far as the 17th century; examples have been found from Persia, China, and Japan. The idea was transformed into a puzzle by the American inventor Sam Loyd in 1858, when he was just 17 years old. Millions were distributed by P. T. Barnum, and versions have shown up in countless books and puzzle kits. It's one of those things that most of us have encountered at some point or other, but not recently—which was perfect grist for the Mel.

Exactly when Mel's insidious variant was developed is not known. At various times he had at least three versions printed up. One rendering, believed to be his most recent, is shown in Figure 1.[1]

Okay, so take a look at the layout. Seem familiar? Sure; you kind of remember this, don't you? The challenge is to cut along the dotted lines, and rearrange the pieces so that each clown is riding a zebra. Right. It's, uh, hey—guess it's not as easy as it appeared.

Indeed, for the simple reason that it can't be done.

Figure 2 shows the Loyd puzzle.[2] Do you see what's different about this compared to the Stover layout?

---

[1] Copyright and marketing rights for these reproductions are retained by the Stover estate.

[2] This particular artwork was used as an advertising premium in the late 1800s; it is reproduced here courtesy of the Slocum collection.

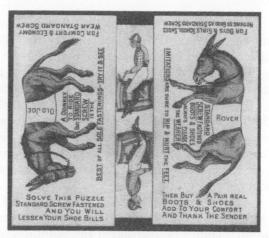

Figure 1.

Figure 2.

Yes, of course, donkeys and jockeys are different from zebras and clowns, but in this context that's merely a cosmetic distinction. The important difference is in the orientation of the quadrupeds. In the Loyd composition, the donkey pieces have identical profiles; that, in turn, enables the solution, as depicted in Figure 3.[3] The card is cut into three pieces. The animals are positioned back-to-back, and then the rider strip is placed crosswise on top to produce the solution.

Figure 3.

---

[3] Another graphic from the Slocum collection, this one used for restaurant promotions circa 1890.

Let's give the Stover layout another perusal. See it now? If you cut the card into three pieces, the zebras will be headed in opposite directions, and Loyd's crosswise solution won't work.

But Mel was, in truth, even sneakier than his victims imagined, for this was a *double* whammy, hiding in plain sight.

The crafty Canadian would always magnanimously hand you *two* of the cards, with the seemingly offhand comment "No need to cut the second one." And you, chomping promptly, interpreted that to mean that the kindly old duffer was providing the two cards for separate purposes: One to slice up and reorganize, the other to keep as an unmarred souvenir.

And that's where he nailed you the second time, because when you reached the teeth-gnashing realization that the pieces of this modified layout couldn't be rearranged successfully, you assumed the sting was over. But in fact, Mel's puzzle *could* be solved, if you used *all* of the materials he'd given you.

Go back and read the text on the zebra card. It's deliberately terse; there's no determinate article in the challenge to "Straddle clowns on two zebras." It's never specified *which* two zebras. So the real solution—the existence of which Mel didn't always bother to mention—is to cut one of the cards along its dotted lines, place one of those zebras back-to-back with its identical counterpart on the whole card, then set the clown strip crosswise over those two zebras to create the desired outcome.

———————

Yet another example of Stoverian stealth was his take on an item of more recent vintage, the pyramid puzzle, which seems to go back less than a century. Despite its relative youth, it has become accepted as one of those venerable conundra that virtually everyone almost remembers. The challenge involves a set of identical pieces that must be assembled to form an equilateral tetrahedron (i.e., a three-dimensional pyramid with four matching triangular sides).

In its best known format, there are two five-sided pieces, as shown in Figure 4. Despite the simplicity of the apparatus, the solution is trickier than one would expect. I won't spoil it by explaining it here; if you want to make a set of pieces with which to experiment, you can use the template in Figure 5 to fold your own.

The arrangement is not easy to remember, as it is rather counterintuitive. Further impeding the task of cerebral retention is the fact that, somewhere along the line, someone came up with the idea of bisecting each piece, thus producing a version wherein *four* identical pieces have to be tetrahedonally assembled.

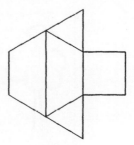

Figure 4.   Figure 5.

This is where Stover stepped in. He made use of the two-piece version, in a common small plastic edition just under an inch and a half wide. That's conveniently small enough to make it an easy endeavor to fingerpalm a third piece. (Perhaps you see where this is going; Mel's victims certainly didn't.)

He'd begin by bringing out the three pieces, using two to form a completed pyramid. As the pieces were so small, his hands hid the details of what he was actually doing. He'd move his hands away, revealing the assembled pyramid while keeping the extra piece concealed in his fingers (Figures 6 and 7).

With his trademark tone of lethargic enthusiasm, he'd remark, "Say, y'ever seen this one?" And, as you were acknowledging that you had, while beginning to ponder the location of that long unvisited mental cranny where the proper configuration was stored, Mel would casually brush his hand against the pyramid, knocking it apart—and, in so doing, he'd add the extra piece, as shown in Figure 8.

You can't get there from here. Welcome to Melville.

These intellectual Chinese Finger Traps had an almost mystical allure; Mel's spells were irrisistible. A further enticement was the distant, ethereal possibility that one might just be able to beat him at his

Figure 6.   Figure 7.   Figure 8.

Figure 9.

own game, by discovering a linguistic loophole or conceptual corner that might afford an alternative solution.

It is with great fondness that I recall the first of the very few times I managed to trump Mel. This momentous event took place in a coffee shop, around 1977. Mel turned over his paper placemat, took out a ball-point pen, and produced a scrawl much like the one shown in Figure 9. This, he contended, was a row of oak trees. The riddle: What number did this picture represent?

Warily, I decided to take the least embarrassing route, and plead ignorance. This, of course, elicited a gleeful cackle from the Manitoba mortifier, who revealed that the answer was nine. "See?" he exclaimed, stabbing his finger onto each oak. "*Tree, tree* and *tree*—that makes nine."

Without pause, he drew a little smudge at the base of each tree as he continued, "A dog comes along, and urinates on each tree. Now what number does it represent?"

Inescapably, I knew that I was going to feel excruciating chagrin at missing what would surely turn out to be a conspicuously obvious answer, but opted to proclaim my continued ignorance. This provoked another self-satisfied snicker as he declared that it was, quite plainly, ninety-nine. He gestured again toward the crude illustration and elucidated: "*Dirty tree, dirty tree* plus *dirty tree*."

He furthered the story, enlarging the smudge at the base of each oak while describing the return of the dog, who this time took the additional effort to defecate on the base of each tree. "Okay," he taunted, "what number is represented now?"

By this time, my dander had been roused from its normally benign and prone posture, and I insisted that Mel wait at least a few moments before charging ahead to announce the solution.

Time stood still, and then, epiphany. "Aha." I crowed. "I've got it! It represents one hundred and eighty-nine."

Irked, my now scowling tormentor said, "No. The answer is one hundred: *Dirty tree and a turd, dirty tree and a turd* plus *dirty tree and a turd.* Now, how the hell did you get a hundred and eighty-nine out of this?"

I leaned back against the curved plastic seat—if memory serves, it was a light orange color not found in nature—and produced my own vainglorious smile, as I explained: "*Shits de tree, shits de tree* and *shits de tree.*"

Life was good.

Mel's last few weeks were spent at the Cedars-Sinai Hospital in Los Angeles. During one of our final meetings, as he lay festooned with tubes and wires, he suddenly brightened, and began the actions of transferring a small object back and forth from hand to hand. Then he stopped, and extended both palm-up fists in front of him.

"Okay," he said, "which hand has the Viagra pill?"

I paused to consider. Slowly, the middle finger of his left hand rose to vertical position.

We all smiled.

Bibliography

[Gar56] Martin Gardner. *Mathematics, Magic and Mystery.* Dover, 1956.

Mel Stover

Martin Gardner

I was greatly saddened to learn of the death in 1999 of my longtime friend Mel Stover, of Winnipeg, Canada. He combined a great love of magic with a sound knowledge of mathematics, chess, checkers, and bridge, all combined with a wonderful sense of humor. One of his earliest contributions to recreational math was his variation of puzzlist Sam Loyd's famous "Get Off the Earth" paradox in which you made a Chinese warrior disappear by rotating a disk. Stover's amusing variations involved switching two parts of a picture to cause a man's face seemingly to change to a glass of beer. (You'll find a picture of the item in my *Mathematics, Magic and Mystery* [Gar56].) Mel had a large collection of such "geometrical vanishes," as they have been called, from which he drew for his article, "The Disappearing Man and Other Geometrical Vanishes" in 1980 [Sto80]. (See Color Plate I.)

A later creation by Mel was another take-off on a Sam Loyd advertising premium known as the "Trick Donkeys." It consisted of three cards, two bearing pictures of a donkey and one with a picture of two riders. The problem was to arrange the cards so that each rider was astride a donkey. Mel's version, which he sold, was fiendishly clever. Each purchaser got two sets of cards. Instead of donkeys, the pictures were of two zebras and two clowns. You were told to cut out the three cards from one set, and keep the other as a spare. The puzzle was absolutely impossible to solve unless you realized that *both* sets of cards had to be used!

Mel and I collected magic tricks and stunts that had a blue or off-color angle. We frequently exchanged such material to keep our collections up

Martin Gardner is the author of some 70 books dealing with mathematics, science, philosophy and literature.

to date. For a brief period Mel contributed very funny items to a short-lived periodical devoted to blue magic—a magazine he himself initiated.

I always looked forward to a letter from Mel because it usually contained a new puzzle, often a puzzle that Mel had invented. Here are four typical brainteasers that first came my way in a letter from Mel:

1. In the equation

$$26 - 63 = 1$$

   change the position of just *one* digit to make the equation correct.

2. Form the figure of a giraffe, as shown below, with matches or toothpicks.

   Change the position of just *one* piece so as to leave the giraffe exactly as it was before, except possibly for a rotation or reflection of the original figure.

3. Tom's mother had four children. Three were girls. The girls' first names were Spring, Summer, and Autumn. What was the first name of the fourth child?

4. A Frenchman who couldn't speak English entered a hardware store. He made sawing motions with his hand. The clerk guessed at once that he wanted to buy a saw.

   An hour later a deaf and dumb man came into the store. He signaled his intention by poking a finger in his left ear, then made circular motions around his other ear. The clerk correctly guessed that he wanted a pencil sharpener.

   The next customer was a blind man. How did he signal to the clerk that he wanted to buy a pair of scissors?

Mel caught me completely off guard with this one. He sent a series of digits broken into parts like so:

XX XX XX XX

I was asked to determine the next pair of digits in the sequence. I couldn't solve it. You can imagine my chagrin when Mel informed me that they were the first eight digits of my home telephone number!

Mel was a personal friend of Howard Lyons, of Toronto, who edited a magic periodical titled *Ibidem*. His artist wife, Pat Patterson, did *Ibidem*'s hilarious silk-screen covers. One of them pictured the dome of the Vatican. In the center of the ceiling a playing card was thumbtacked. A visiting magician had performed the classic card on the ceiling trick. Pat was also responsible for drawing the leprechauns in a popular marketed version of a geometrical vanish.

On a summer cruise that Mel took with Howard and Pat, Mel won first prize on the ship's costume ball. His "costume" consisted of a huge cardboard hat that Pat made to cover his entire head. On Mel's bare chest and abdomen Pat painted a face. Mel's nipples were the eyes. His navel was the nose, and under it Pat painted a wide red mouth. I thought I still owned a great photograph of Mel in this outlandish get-up, but I was unable to locate it for this article.

After Mel's wife Mary died, he attended almost every magic convention held in the United States, Canada, or abroad, even after an accident confined him to a wheel chair. I never met anyone who didn't like Mel and thoroughly enjoy his company. In his elderly years he contributed a column called Brainteasers to *Genii,* a leading magic monthly. In the September 1999 issue of this magazine, you'll find a splendid tribute to Mel by his friend Max Maven, a well known magician specializing in what the trade calls "mental magic" [Mav99]. The article includes a cardboard insert picturing Mel's clown and zebra puzzle.

Here are the answers to the four problems I gave:

1. $2^6 - 63 = 1$.

2.

3. Tom.

4. The blind man said, "I want to buy a pair of scissors."

Bibliography

[Gar56] Martin Gardner. *Mathematics, Magic and Mystery.* Dover, 1956.

[Mav99] Max Maven. Just for the mel of it. *Genii*, September 1999.

[Sto80] Mel Stover. The disappearing man and other geometrical vanishes. *Games Magazine*, November–December 1980.

The Stack of Quarters

Ronald A. Wohl

To Mel

This contribution is dedicated to Mel Stover. I had known Mel for many years and attended more magic gatherings with him than I care to enumerate. Mel truly was brilliant and Mel and I had many wonderful conversations.

And who could ever forget, after witnessing it at G4G2, the electric cart race between Mel and J. C. Doty, from the hotel up and down Peachtree Street to the 191 Club. Truly, the Atlanta 2 were so much more exciting than the Indianapolis 500!

We all miss Mel, both as person and as inspiration, and we all missed Mel at G4G4. But perhaps, just perhaps, Mel had the last laugh. I somehow cannot get rid of the feeling that Mel attended G4G4, albeit in the 4th dimension. While many of the participants made wonderful attempts to explain, illustrate and cajole the 4th dimension, and to make it appear very, very real, there may yet be some aspects of the 4th dimension that we have not fully grasped yet, but that Mel now knows.

And Mel and I are looking forward to G4G5!

The Stack of Quarters

The following is a wonderful "bar stunt," or puzzle for all occasions. It is not mine, and I have been told that this presentation is by Mel Stover.

Ronald A. Wohl's 50-year involvement in magic and mathematics is documented in his articles in magic magazines in Switzerland, Germany, England, and the United States, and in Martin Gardner's books.

A stack of 10 quarters is placed on the table. Also placed on the table, in a row, are all the standard American coins, in order from the largest coin to the smallest coin: An Eisenhower (or similar) dollar, a Kennedy (or similar) half-dollar, a Susan B. Anthony or the new golden dollar coin, a quarter (any state will do), a nickel, a penny, and finally a dime. (Yes, in case you did not know, the penny is larger than the dime.)

All spectators are invited to guess which coin, when placed vertically besides the stack of 10 quarters, will match its height perfectly. Each spectator is asked in turn, in order to commit himself, by pushing the coin he has selected in his mind forward an inch or so.

The majority of people will choose a coin that is way too large. Very rarely will the correct coin be chosen, which is—**suspense!**—the smallest coin, the dime, as can be easily demonstrated once all spectators have made their selection.

A Few Tips for Presentation

It is nice to have all the necessary coins together in a special coin purse.

The illusion is heightened if the stack of quarters is not placed directly onto the table, but onto a support such as a matchbox or a deck of cards.

Do not assemble the stack in your hands, but build it up slowly and painfully, one by one, while counting loudly from one to ten.

Do not place the row of coins too close to the stack, but reasonably far away. It seems best, if the row of coins extends to both sides of the stack (when drawing a perpendicular line, etc.).

And finally, for magicians only: After everybody has digested this puzzle, the perfect moment has come to perform a trick with a "stack" of 10 quarters. You could not ask for better misdirection!

Part II
Tantalizing Appetizers: Challenges for the Reader

Crostic in Honor of Martin Gardner

Julie Sussman

Instructions

When you have solved a crostic:

- the letters in the diagram (reading across) will spell out a quotation;

- the first letters of the answer WORDs (reading down) will spell out the author and title of the work from which the quotation comes.

Just write the answers to any CLUEs you know over the numbered dashes next to the CLUEs, then transfer these answer letters to the correspondingly-numbered diagram squares. For example, if the CLUE and answer to WORD B were

B Crostic $\underset{4}{\text{P}}$ $\underset{13}{\text{U}}$ $\underset{2}{\text{Z}}$ $\underset{21}{\text{Z}}$ $\underset{19}{\text{L}}$ $\underset{48}{\text{E}}$

you would place a P in square 4, a U in square 13, etc.

You can also work backwards from the diagram to the WORDs, since a letter in the corner of each square tells which answer WORD it comes from. For example, if your diagram contains:

48 B	49 X	50 C	51 P
E	A		Y

Julie Sussman is the author of *I Can Read That! A Traveler's Guide to Chinese Characters*, and has coauthored and edited computer science textbooks for M.I.T.

you might deduce that square 50 should contain an S. This S should be placed over dash 50 in WORD C.

Have fun! And be careful when reading the diagram—only black squares (not the end of the line) separate words.

			1 Z	2 V	3 A		4 O	5 J	6 Y		7 D	8 G	9 J	10 C	11 E
12 N	13 V	14 H	15 B		16 J	17 C	18 E	19 G	20 N	21 B	22 H	23 T		24 D	25 F	26 E	27 N	28 J
29 G	30 R	31 L		32 C	33 P		34 W	35 N	36 V	37 L	38 X		39 S	40 P		41 K	42 O
43 F		44 V	45 N	46 Z	47 J	48 D		49 Q	50 H	51 E	52 T	53 G	54 J	55 N	56 B		57 V	58 D
59 A		60 E	61 W	62 Q	63 M		64 O	65 Z		66 V	67 O	68 S		69 J	70 T	71 C	72 Q	73 N
74 R	75 E	76 B		77 J	78 G	79 N	80 T	81 E	82 F		83 Z	84 S		85 L	86 N	87 V	88 Y	89 E
90 C		91 P	92 X		93 Q		94 N	95 B	96 E	97 J	98 U	99 C	100 G		101 M	102 A		103 F
104 O	105 M		106 B	107 E	108 D	109 N	110 T	111 C	112 F	113 J	114 H	115 G	116 R		117 B	118 Y	119 U
120 X		121 Q	122 G	123 R	124 H	125 N	126 E	127 J	128 L	129 V	130 B	131 C	132 Z	133 T		134 V	135 Z	136 L
137 C		138 G	139 W	140 X	141 B	142 Z		143 E	144 J	145 N	146 G	147 B	148 C	149 R	150 D		151 I	152 Y
153 P		154 X		155 W	156 E	157 R	158 G	159 B	160 N		161 F	162 Z	163 E	164 Q	165 Y		166 C	167 N
168 G	169 E	170 T	171 H	172 B	173 R	174 J		175 Z	176 M		177 C	178 U	179 Y		180 R	181 G	182 Q	183 W
184 X		185 E	186 G	187 B	188 L	189 T	190 N	191 J	192 D		193 Y	194 S		195 H	196 O		197 T	198 G
199 E	200 D	201 B	202 H	203 C	204 N	205 R	206 L	207 J		208 D	209 I		210 P		211 T	212 N	213 A	214 L
215 E	216 H	217 J	218 U		219 N	220 U	221 S	222 F	223 T		224 P	225 K		226 S	227 A		228 W	229 A
230 K		231 I	232 H	233 W		234 U	235 W	236 R	237 X	238 F		239 H	240 J	241 N	242 U	243 L	244 R
245 V	246 N	247 J	248 F	249 T	250 Z		251 I		252 U	253 V	254 X	255 Q					

CLUES   WORDS

A   Beverage featured in Bach
    cantata, BWV.211
    213 229 102 227   3   59

B   Martin Gardner's edition of
    Carroll work (2 wds preceded   95 201 15 21 130 141 172 56 76 117
    by "The")
    159 106 187 147

C   Ancient Egyptian document
    showing "Russian peasant   148 137 131 111 90 71 177 166 17 99
    method" of multiplication
    (2 wds)   10 203

D Tibetan dog (2 wds)

 __48__ __58__ __108__ __150__ __7__ __208__ __24__ __192__ __200__

E Magazine for which Martin
   Gardner started writing in
   the 1950s (2 wds)

 __169__ __126__ __51__ __75__ __18__ __199__ __96__ __60__ __156__ __11__

 __185__ __107__ __89__ __26__ __163__ __143__ __215__ __81__

F Old instrument for observing
   positions of celestial
   bodies

 __112__ __82__ __103__ __238__ __25__ __222__ __43__ __161__ __248__

G Bestseller by word J (3 wds)

 __158__ __115__ __19__ __168__ __78__ __53__ __198__ __146__ __138__ __100__

 __122__ __8__ __29__ __186__ __181__

H "The Mass is ... an ____
   ..." (Martin Luther)

 __232__ __239__ __124__ __22__ __114__ __171__ __195__ __216__ __202__ __14__

 __50__

I Author of humorous poetry

 __151__ __251__ __209__ __231__

J Martin Gardner's successor
   at word E (full name)

 __9__ __144__ __127__ __207__ __97__ __77__ __16__ __28__ __247__ __217__

 __5__ __69__ __113__ __47__ __174__ __191__ __240__ __54__

K Huge Australian bird
   (world's second largest)

 __225__ __41__ __230__

L Gamow's visitor to
   Wonderland (full name)

 __85__ __243__ __136__ __188__ __37__ __128__ __214__ __32__ __206__ __31__

M Portent

 __101__ __63__ __105__ __176__

N Martin Gardner's column in
   "Skeptical Inquirer" (3 wds
   + compound)

 __145__ __73__ __12__ __160__ __55__ __212__ __94__ __86__ __241__ __167__

 __79__ __190__ __109__ __246__ __219__ __45__ __27__ __125__ __204__

 __35__ __20__

O "The ____ of Peers,
   throughout the war, did
   nothing in particular, and
   did it very well" (Gilbert
   and Sullivan, "Iolanthe")

 __104__ __64__ __4__ __196__ __42__

P Author of "Flatland"

 __210__ __91__ __224__ __40__ __153__ __33__

Q Advantage (2 wds)

 __49__ __121__ __72__ __182__ __62__ __67__ __93__ __164__ __255__

R Largest two-digit prime
   (compound)

 __116__ __205__ __149__ __123__ __180__ __30__ __74__ __173__ __236__ __244__

 __157__

S Disposable handkerchief

 __39__ __226__ __84__ __194__ __221__ __68__

T   Appropriately named street Martin Gardner used to live on (2 wds)

170   70   211   249   110   223   197   52   23   133

189   80

U   Gold coin worth 16 pieces of eight

119   218   98   252   178   242   220   234

V   Late character in word B (2 wds)

134   2   13   57   36   87   129   245   44   253

66

W   Homeric hero

61   233   228   155   34   235   139   183

X   "Powers of Ten" collaborator (full name)

254   120   92   184   154   140   237   38

Y   Kind of glass featured in word B

179   6   152   88   193   118   165

Z   "The difference between the first- and second-best things in art absolutely seems to escape verbal _____ ..." (William James)

250   162   65   135   46   83   1   175   132   142

A Clock Puzzle

Andy Latto

The Puzzle

I have a clock on which the hour hand and the minute hand are identical.
Usually this doesn't bother me. If one hand points to 3, and the other
to 12, I know it is 3:00. It can't be 12:15, because at 12:15, the hour
hand wouldn't point exactly to 12; it would point a little bit past the 12.

Assume that I can always tell whether it is A.M. or P.M. Are there
times during the day when I cannot tell the time by looking at my clock?
How many times does this occur in a 12-hour period?

Solution

Imagine two clocks, one normal, and running 12 times faster than the
first. The crucial property the clocks have is that if both started at
12:00, the hour hand of the second clock is always in the same position
as the minute hand of the first. If at time t, the minute hand of the
second clock happens to be in the same position as the hour hand of the
original clock, then I cannot tell which is the minute hand and which
is the hour hand at time t, because I do not know whether the current
time is the time on the first clock, or the time on the second clock. At
time t, the two hands are in the same two positions on the two clocks,
only interchanged. So the difficult question "How many times can I not
distinguish the hands on the first clock?" has been transformed into the
simple question "How often is the minute hand of the second clock in

---

Andy Latto has previously published in *Inside Backgammon* and *The Intelligent Gambler,* and won the 1999 New England Poker Classic.

the same position as the hour hand of the original clock?" Over a 12 hour period, the hour hand of the original clock revolves once, while the minute hand of the fast clock revolves 144 times, so they are in the same location 143 times. So, 143 is the answer to the question "How often am I unable to tell which hand of my clock is the hour hand?"

However, this is subtly different from the original question, "How often am I unable to determine the correct time from my clock?" At times when the two hands of my clock are at the same location, I cannot tell which is the hour hand and which is the minute hand, but I can nonetheless tell what time it is. Therefore, starting with the 143 times that we cannot distinguish the hands of the clock, we must subtract the 11 times when we cannot distinguish the hands, but do not need to in order to determine the time, to conclude that there are 132 times in a day when my clock will not tell me the time.

It is interesting to note that this is essentially a topological result. The above solution shows that even if my clock runs erratically (with the hands not running steadily nor in sync), as long as

1. both hands move constantly forward throughout the day,

2. the hour hand never passes the minute hand, and

3. the minute hand revolves 12 times while the hour hand revolves once, then

there will be exactly 132 times during the day when I cannot tell what time it is.

Three Problems

Andy Liu and Bill Sands

Problem 1 *What is the maximum number of non-overlapping circles of radius 1 which can simultaneously have exactly one point in common with a square of side length 2?*

Problem 2 *Let x and y be positive integers each with exactly two kinds of digits. For example, $x = 1313$ and $y = 3344$. Then $x + y = 4657$ has four kinds of digits. However, this is far from optimal.*

    (a) What is the maximum number of kinds of digits in $x + y$? Give an example in which the optimal result is attained, using the minimum number of digits in x, y, and $x + y$ combined, and give a justification.

    (b) Answer the same question for $|x - y|$ instead of $x + y$.

Problem 3 *Dissect the figure below into three pieces and reassemble them into an equilateral triangle.*

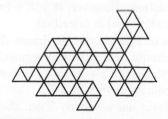

Andy Liu has won international and national awards for the promotion of mathematics. **Bill Sands** likes winters on the Canadian prairie, and was an editor of *Crux Mathematicorum* for ten years, but is otherwise relatively sane.

Comments. Problem 1, proposed by Bill, has since appeared as Challenge Problem C33 in the May 1998 issue of *Mathematics and Informatics Quarterly.* Problem 2(a) was proposed by Bill and has since appeared as Problem S-14 in the April 1998 issue of *Math Horizons.* The current version was a modification by Andy, who also extended it to Problem 2(b), appearing here for the first time. Problem 3 appeared as Problem 1806 in the January 1993 issue of *Crux Mathematicorum,* of which Bill was then editor. It was proposed by Andy, based on an idea of NOB Yoshigahara.

Solutions

Problem 1   We can easily have six such circles. We now prove that we cannot have seven. The following diagram shows the locus $\mathcal{L}$ of a point at a distance 1 from the given square S. It consists of four line segments $BC, DE, FG,$ and HA obtained by translating the edges of S outward, together with four circular quadrants $AB, CD, EF,$ and GH connecting them.

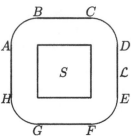

Suppose it is possible for seven non-overlapping unit circles to touch S simultaneously. Then the centers of these circles determine a convex heptagon T inscribed in $\mathcal{L}$. Moreover, each side of T is of length at least 2, so that its perimeter is at least 14. If the length of $\mathcal{L}$ is less than 14, we will have a contradiction. However, it is $8 + 2\pi > 14$. So we seek a shorter curve $\mathcal{L}'$ in which T is still inscribed.

Suppose no vertices of T lie on the quadrants AB and CD. Then we can replace these quadrants in $\mathcal{L}$ with the segments AB and CD. The new curve $\mathcal{L}'$ has length $8 + 2\sqrt{2} + \pi < 14$. Hence at least three of the quadrants in $\mathcal{L}$ must contain a vertex of T.

Suppose each quadrant has a vertex. Then there is a segment without. However, we can move onto this segment a vertex on an adjacent quadrant without causing any problem. Thus we may assume that the vertices of T lie on BC, CD, DE, EF, FG, GH and HA respectively.

Take U on HA and V on CB such that $AU = BV$ and $UV = 2$. Then either AU or BV does not contain a vertex of T. We may assume that AU does not. We can then replace the quadrant AB and the segment UA in $\mathcal{L}$ by the segment BU. The saving amounts to $\sqrt{2} - 1 + \frac{\pi}{2} - \sqrt{3}$.

Let P be the vertex of T on the quadrant CD. We can replace this quadrant with the segments CP and DP. Now $CP + DP \leq CM + DM$, where M is the midpoint of the quadrant CD. Then the saving amounts to $\frac{\pi}{2} - (CP + DP) \geq \frac{\pi}{2} - 2\sqrt{2 - \sqrt{2}}$.

Let $\mathcal{L}'$ be obtained from $\mathcal{L}$ by replacing the quadrants AB and CD as indicated above, and the quadrants EF and GH in the same way as for CD. Then its length is at most

$$8 + 2\pi - \left(\sqrt{2} - 1 + \frac{\pi}{2} - \sqrt{3}\right) - 3\left(\frac{\pi}{2} - 2\sqrt{2 - \sqrt{2}}\right) = 9 + 6\sqrt{2 - \sqrt{2}} + \sqrt{3} - \sqrt{2} < 14.$$

This is the desired contradiction.

Comments. For completeness, we give a proof of the following result which was used in the solution above.

Lemma. If M is the midpoint of a circular arc CD, then $CP + DP < CM + DM$ for any other point P on this arc.

Proof. We may take P to be on the arc DM. Extend CM to N so that $DM = MN$ and extend CP to Q so that $DP = PQ$. Since $\angle CMD = \angle CPD$, we have $\angle CND = \angle CQD$. Hence C, D, N and Q are also concyclic, and the centre of this circle is M. Hence $CM + DM = CN > CQ = CP + DP$. $\qquad\square$

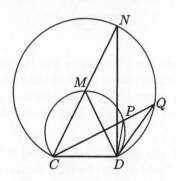

Problem 2

(a) A solution with all 10 digits is $4949994494 + 222244 = 4950216738$. We now prove that 26 digits in all is minimum. Suppose to the contrary there is a 25-digit solution. First assume that $x + y$ has exactly 10 digits. Then x has at least 9 digits. Suppose x has 9 digits, so that y has 6. Then the first 3 digits of $x + y$ will be 100, and it will not have 10 distinct digits. Suppose x has 10 digits, so that y has 5. If the carrying does not reach the third digit of $x + y$, then its first 3 digits are not distinct. If it does, then the fourth and fifth digit of $x + y$ are both 0. In either case, $x + y$ will not have 10 distinct digits. If $x + y$ has more than 10 digits, the situation is worse, and an analogous argument shows that it cannot have all 10 kinds of digits.

(b) A solution with all 10 digits is $4040004004 - 442222 = 4039561782$, and a similar argument shows that 26 digits in all is minimum.

Remark. There is a striking affinity between the solutions to the two parts, with 4 appearing in x as well as y, and 2 in y.

Problem 3

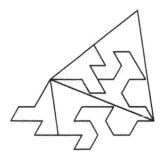

Remark. Note that both the given shape and the equilateral triangle tile the plane. Thus the dissection may be obtained by superimposing two tessellations on each other. This is a standard method employed by Harry Lindgren in [Lin72]. See also Greg Frederickson's [Fre97].

Comments. Problem 1 was solved by Andy and Bill. Problem 2(a) was solved by Andy, and Problem 2(b) by Nick Baxter of Hillsborough, California. Problem 3 was solved by P. Penning of Delft, The Netherlands.

Bibliography

[Fre97] Greg N. Frederickson. *Dissections: Plane and Fancy.* Cambridge University Press, New York, 1997.

[Lin72] Harry Lindgren. *Recreational Problems in Geometric Dissections and How to Solve Them.* Dover Publications, New York, 1972. Published in 1964 under title *Geometric Dissections.*

A Scrub Tile Puzzle

Tom Rodgers

Paste the following 3-letter words on ten one inch square tiles:

CAR CUB DIM HEN HUT MOB RED SAW SON WIT

Note that every letter occurs exactly twice and that any two letters occur in at most one word. These are called *scrub tiles* and operate like word dominoes in that two scrubs can abut only if they share a letter in common. For this puzzle, you'll want to cut out the 10 Scrub tiles from heavy cardboard, and construct a board with 10 squares:

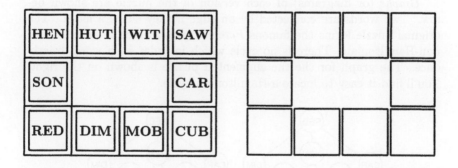

See if it is possible to place the 10 scrub tiles on the board. (As drawn, they are almost placed legally, except that **SON** and **RED** abut despite sharing no letters in common.)

Tom Rodgers organized the four Gatherings for Gardner and collects puzzles.

Solution

Before reading on, verify for yourself that the following solves the puzzle:

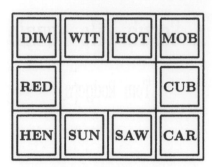

The astute reader will notice an error. Indeed, any two words which abut share a letter. However, the scrub tiles aren't identical to those in the original puzzle! We've replaced **SON** and **HUT** by **SUN** and **HOT**. In truth, the original problem has no solution.

Make a copy of all 12 scrubs and the board out of heavy cardboard. Find a likely victim, and arrange your scrubs on the grid as shown in the "solution," palming the two scrubs, **HUT** and **SON**. Briefly show your mark the arrangement and then drop the scrubs on the table. Be careful to hold onto the **HOT** and **SUN** scrubs with your thumbs, letting **HUT** and **SON** fall in their place. Your poor, befuddled victim will have no chance at finding a loop.

Graphs (or diagrams) of each version of the puzzle are shown below. Two words are connected by an edge if they share a letter. The original puzzle forms the famous *Petersen graph* which is known to be non-Hamiltonian: There is no cycle which includes each word exactly once. The graph for the the augmented puzzle is shown on the right. You'll find it easy to locate a Hamiltonian cycle.

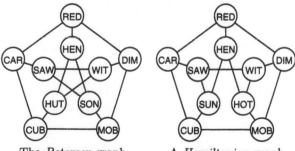

The *Petersen* graph   A *Hamiltonian* graph

See Martin Gardner's articles about Snarks for other interesting properties of the Petersen Graph [Gar76]. For more information on scrubs, see [CFR99].

Bibliography

[CFR99] Zhi-Hong Chen, Jeremiah Farrell, and Thomas Rodgers. Scrubwoman Edith meets W. R. Hamilton. *Word Ways: The Journal of Recreational Linguistics*, 31(1):69, February 1999.

[Gar76] Martin Gardner. Mathematical games. *Scientific American*, 234(4):126–30, April 1976.

All Tied up in Naughts

David Wolfe and Susan Hirshberg

It's your turn to play in each of these 2-player Scrabble® positions. Find the play which yields the best chance to win the game.[1]

The avid Scrabble® player should try the problems now before reading on. Anyone else will surely want to read on for a hint or two; we assure you the hints will not give away the problem.

Recall that there is a bonus of 50 points for placing all 7 tiles in one turn (called a *BINGO*). Any tiles remaining in your rack at the end of the game are subtracted from your score. If, however, one player ends the game by using up all his/her tiles, that player adds double the total value of tiles in the opponents rack to his/her score. A tile on the board with a lowercase letter is a blank.

All words on each board are legal. In addition, the solver may also need to know the following unusual words: AA, AY, ALDRINS, BA, DARINGS, DURIANS, EN, ES, GRADINS, ILEXES, MU, NA, OS, PED, QINDARS, REFT, RIBANDS, UN, YOND, YONI.

Hint: Don't go for points on your first play! (Another hint is in the first paragraph of the solution page.)

David Wolfe is best known for his work with Elwyn Berlekamp on mathematics applied to the game of Go. **Susan Hirshberg** is conspicuously obscure.

[1] These problems are based on the American lexicon in any edition of *The Official Scrabble® Players Dictionary*. (For Scrabble® enthusiasts, use any lexicon you wish from OSPD up through TWL98.)

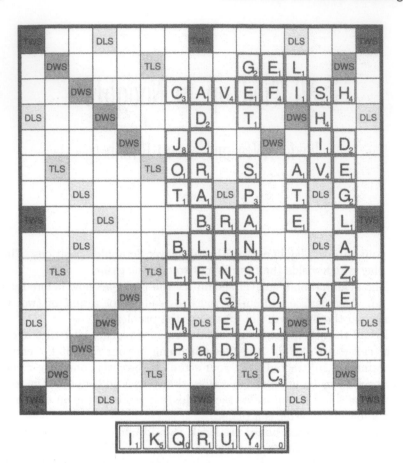

Tiles Left: AAAEEEFIIIMNNNNOOOOORRRTTUUUWWX

Figure 1.   The score is tied. What is your best plan to win?

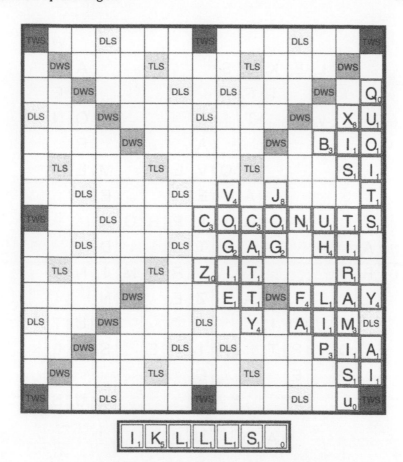

Tiles Left: AAAAABDDDDDEEEEEEEEEEEEFGHIMN
NNNNOOOOOPRRRRRTTUUVWW

Figure 2. The score is tied. What is your best plan to win?

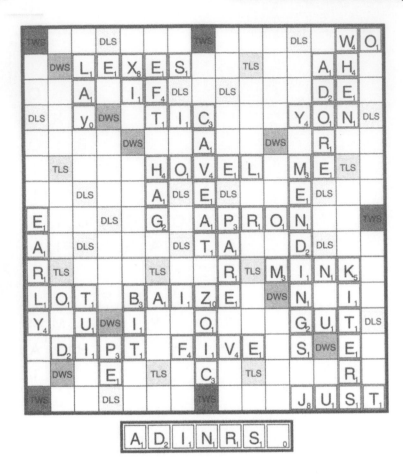

There are 9 tiles remaining: BGLNOOQUW

Figure 3.   The score is tied. What is your best plan to win?

Solutions

Surprisingly, the best play in each instance is to place the blank tile in your rack next to the blank tile already on the board, scoring no points for the turn. Before reading on, the puzzle solver may want to rework the problems.

Figure 1. For this problem, there are two high point plays that come to mind: QUIcK/SUAVE/PIT/BRAcE/BLINK (64 points) and QUIRKs/BLIMPs (57 points). Although playing the blank as an "s" below the "a" in PaDDIES scores no points this turn, it guarantees you a 111 point play next turn of QUIRKY/asK. Notice that the opponent cannot have an S, P, H or K for ASS, ASP, ASH and ASK, and has no way of blocking the QUIRKY opening.

Figure 2. SkILLfuL/AIL (149 points) looks awfully good, but playing the blank for xu (0 points) is superior. The board is then so blocked up that the game can only last a few more turns. The next turn, you plan to play ZITS/JOGS (25 points) which the opponent cannot block. In the next one or two turns, you can exchange so as to guarantee your rack has at worst three 4 point tiles and four 1 point tiles, which could be 9 points worse than opponents rack. In the meanwhile, the best the opponent can hope for is to play WON/JO (15). (If he tries playing JO/ON alone, you can play ION.) Thus this 0 point play guarantees a win by at least $25 - 15 - 9 = 1$ point.

It's important that the opponent can have no L's (for ZLOTE/LET), Y's (for ZYME/YET), or S's (for lots of reasons), since these plays threaten to prolong the game. Also, the Scrabble player well versed in the American lexicon can verify there is no other way to thoroughly block up the lower right corner and guarantee a quick ending to the game. For example, guL might be extended to MOguL, Sum to OMASum, and puL to AMpuL.

Figure 3. For this last problem, it is possible to *bingo* and play all 7 tiles in any one of a number of ways. There is a chance, however, that the opponent will bingo back with BLOWGUN/PEW or LONGBOW/PEG (along the bottom) or LONGBOW/LEFT/OS (along the top), leaving you behind, since you'll be stuck with the Q. How can you guarantee a win?

If you stall for a turn by passing, the opponent might play the MU/UN (12 points), and you may still lose for the same reason as above.

Your only hope to guarantee a win is to play only one tile. But where? If you were to pick up the Q, your only bingo is then QINDARS, and the only opening is along the bottom (QINDARS/PEN). If the opponent blocked it with LONGBOW or BLOWGUN, you could lose.

To create a second possible place to play QINDARS, play the blank next to the other blank to form "ye" (0 points). This guarantees you can bingo next turn no matter what your adversary does. (If, by the way, the opponent passes, you bingo if you pick up the Q, accumulating enough points to win; you play OW/AWE if you pick up a W and bingo next turn; and you can safely pass if you pick up any other tile, leaving you with fewer points in your rack.) The following table shows how you have two possible places to bingo no matter what single tile you draw; hence the opponent cannot block both:

		1	2	3	4
B	RIBANDS	X		X	
G	DARINGS		X		X
	GRADINS			X	
L	ALDRINS	*		X	
N	INNARDS		X	X	
	INROADS	X	X		
O	ORDAINS			X	
	SADIRON				
Q	QINDARS	*		X	
U	DURIANS		X		X
W	INWARDS	*	X	X	
-	BLOWGUN			X	
-	LONGBOW			X	X

The four bingo lanes are numbered:

1. Upper left (ILEXES or AA)
2. Right side (YOND or YONI to MINKS)
3. Lower side (PEA, PED, PEG, PER or PEW)
4. Upper side (DEFT, LEFT or REFT)

"X" means the bingo can be played in the lane.
"*" means the bingo can be played after playing "ye".

Jumping Cards

Jaime Poniachik

> Playing cards, with their numerical values, have long provided recreational mathematicians with a paradise of possibilities.
> —Martin Gardner

Before describing my jumping card problems, a few words about Martin Gardner are in order. Martin Gardner is the founding father of the mathematical recreations world in Argentina. Through his columns and books (the *Canon*) we got to know both the classics—like Sam Loyd and H. E. Dudeney—and the contemporaries. It is because of Gardner that we Argentinian mathemagicians even got to know one another. Here is one such example: In my first letter to Martin I told him that math puzzlers in Buenos Aires were few and isolated. In his answer he sent me the name and address of a Federico Fink, a few blocks from my home. Federico, who was a keen aficionado of Mac Mahon's dominoes, soon became a close friend of mine. Martin taught us the puzzles and also the spirit and ethics of puzzling. We love him.

Close Neighbors

For the first problem, begin with cards 1 to 8 in ascending order:

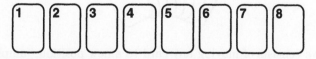

Jaime Poniachik is a well-known maker of games and puzzles. He leads Argentina's foremost publishing company of puzzle books and magazines.

The object of the puzzle is to move the cards into descending order through a sequence of legal moves:

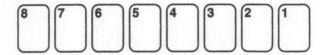

A legal move consists of exchanging any two cards with one condition: two neighboring cards may never have a difference strictly greater than 3. On a first move we could not exchange 1 and 4, because it would leave a pair of neighbours, 1 and 5, with a difference exceeding 3:

1. How many moves do you need to do the task?

2. Try a similar puzzle with cards 1 to 9.

Going over Multiples

Let's try a new rule. We begin again with cards 1 to 8 in ascending order. This time it is convenient to hold the cards in your hand:

The object of the puzzle is to reverse the order:

Each move consists of moving any card to a new position, jumping over one or more cards, under one condition: you may only jump over cards which numbers add up to an even result.

On a first move you could take card 7 over $6 + 5 + 4 + 3 (= 18)$ and insert it between 2 and 3.

1. How many moves do you need to complete the task?

2. A more demanding puzzle: same, but you may now only jump over cards which numbers add up to 5 or a multiple of 5 (5, 10, 15, 20, 25, ...).

Measured Jumps

Let us still try a new rule. We hold in hand cards 1-2-3, in ascending order:

The object of the puzzle is to reverse the order. Each move consists in taking card number N over precisely N other cards. The diagram shows a solution in 4 moves:

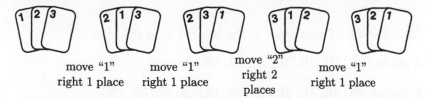

| move "1" | move "1" | move "2" | move "1" |
| right 1 place | right 1 place | right 2 places | right 1 place |

Though the example shows cards moving to the right only, you may make some moves toward the right and other moves toward the left at your convenience. Try to solve the same puzzle with differing numbers of cards. The following chart shows the best we could do.

Starting	Ending	Moves
1	1	0
12	21	1
123	321	4
1234	4321	7
12345	54321	11
123456	654321	14
1234567	7654321	18
12345678	87654321	21
123456789	987654321	25

Answers

Close Neighbors

1. 14 moves. These are the cards to exchange on each move: 13, 76, 24, 73, 15, 74, 84, 64, 23, 56, 12, 35, 31, 24. (Solver: Héctor San Segundo)

2. 24 moves: 97, 68, 34, 37, 95, 65, 53, 54, 24, 56, 97, 32, 87, 31, 67, 53, 54, 87, 64, 56, 98, 75, 68, 97. (Solver: Diego Bracamonte)

Going over Multiples

1. 7 moves: 5 over 4321, 1 over 23467, 8 over 1764325, 6 over 4325, 2 over 347, 7 over 4356, 3 over 4.

2. 9 moves: 8 over 7+6+5+4+3, 6 over 5+4+3+8, 7 over 5+4+3+8, 1 over 2+6+7+8+3+4, 8 over 7+6+2, 2 over 6+7+3+4, 6 over 7+3, 3 over 6+4, 5 over 1+2+3+4. (Solvers: Nina Poniachik, Héctor San Segundo)

Measured Jumps   Cards move to the R(ight) or to the L(eft).

4 cards: 2R, 1R, 3R, 2L, 1R, 1R, 2R.

5 cards: 2R, 3R, 1R, 3L, 2L, 1R, 4R, 1R, 2R, 3R, 2R.

6 cards: 3R, 2R, 4R, 1R, 4L, 3L, 2L, 1R, 5R, 1R, 4R, 2R, 2R, 3R.

7 cards: 3R, 4R, 2R, 5R, 1R, 5L, 4L, 3L, 1R, 1R, 6R, 1R, 5R, 3R, 2R, 2R, 4R, 3R.

8 cards: 4R, 3R, 5R, 2R, 6R, 1R, 7R, 4L, 5L, 6L, 2L, 1R, 1R, 1R, 6R, 3R, 5R, 4R, 2R, 2R, 2R. (Solver: Rubén Efron)

9 cards: 4R, 5R, 3R, 6R, 2R, 7R, 1R, 8R, 5L, 6L, 7L, 2R, 2R, 7R, 1R (repeat six times), 6R, 4R, 5R, 3R, 3R. (Solver: Rubén Efron)

This puzzle may be solved even if you are obliged to move every card towards the right. In fact, if on each turn, you move the card of the highest value allowable towards the right, the procedure will terminate with the task complete.

Six Off-beat Chess Problems

John Beasley

The chess problem is one of the most specialized forms of puzzle, and it occurs in Martin Gardner's books only where there is some twist requiring a leap of imagination rather than mere grandmasterly depth of calculation. The six problems given here are all of this kind. In each case, the stipulation is the apparently elementary

> **White to play and force mate in two moves against any defense**

but naturally there is a catch, and to get around the catch the solver must first answer the question

> **What was Black's last move?**

Sometimes the catch will be found not to be a catch at all; sometimes it is all too real, but if you cannot get in by the door then perhaps the window will be open ...

Figure 1.

Figure 2.

John Beasley, the author of *The Ins and Outs of Peg Solitaire*, is a well-known composer of chess problems and endgame studies.

65

Figure 3.

Figure 4.

Figure 5.

Figure 6.

Solutions

Figure 1. (F. Amelung, *Duna Zeitung*, 1897) A mate in two appears quite impossible, but what was Black's last move? Not by Ph7, which has never moved at all; not by K from g6, because the two kings would have been on adjacent squares; not K from g7, because it would have been in check to white Pf6, and there is no square from which the pawn could have come to give this check; not by Pg5 from g6, because there it would have been giving check to white K with Black to move. So Black's last move must have been g7-g5, and White can capture it *en passant*: **1 h5×g6** (1 ... Kh5 2 R×h7 mate). Amelung (1842–1909) was a famous Latvian chess endgame analyst and problem composer.

Figure 2.   (S. Loyd, *Musical World*, 1859)
1 Qa1 would threaten mate by 2 Qh8, and
Black's only defense would be to castle. Can
he do so? Well, Black's last move cannot
have been with a pawn, since neither of his
pawns has moved, so it must have been with
the king or the rook. Hence he has lost his
right to castle, and **1 Qa1** does indeed force
mate in two. Sam Loyd (1841–1911) was
America's most famous puzzlist, whose work
has been featured in Martin's books right
from the start.

Figure 3.   (Simplified from a problem by S.
Loyd, *Missouri Democrat*, 1859) Again, can
Black castle? If his last move was by the
rook on a8, he has lost the right to castle
long, and **1 Qg7** forces mate in two; if it was
by the rook on h8, he has lost the right to
castle short, and **1 Q×c7** works. (If his last
move was by the king, he has lost the right
to castle on either side, and either Qg7 or
Q×c7 will work.) So White always can force
mate in two from this position, but precisely
how he does it depends on Black's last move.
Loyd thought this idea too slight to show in
simple two-move form, and buried it inside
a three-mover.

Figure 4.   (W. Langstaff, *Chess Amateur*
1922) A combination of these two themes.
Either Black's last move was with his king
or rook, in which case he has lost the right
to castle and **1 Ke6** forces mate in two,
or it was with the pawn, in which case it
must have been Pg7-g5 and the *en passant* capture **1 Ph5×g6** works (1 ... 0-0 2
h7). Again, White definitely can force mate
in two, though how he does it depends on
Black's last move. Langstaff (1897–1974)
was a British chess problem composer, and
this little classic has kept his name alive.

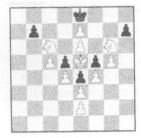

Figure 5.   (T. R. Dawson, *Falkirk Herald*, 1914) More of the same? Black's last move cannot have been with K (for example, on d8 it would have been in an impossible double check from Nc6 and Pe7), so it must have been Pd7-d5 or Pf7-f5 and White takes en passant as appropriate. No! White has made ten pawn captures (a×b, b×c twice, c×d twice, d×e twice, f×e, g×f, and h×g) and Black has lost only ten men including the light-square bishop, so Black's d-pawn must have moved long ago to let this bishop out to be captured. So Black's last move must have been Pf7-f5, and the solution is explicitly **1 Pg5×f6**. Dawson (1889–1951) was Britain's most famous chess problem composer and delighted in tricks of this sort.

Figure 6.   (E. Dunsany, *Fairy Chess Review* 1943) Again, more of the same? It can be shown that Black Bb1 is Black's original Bc8, so his move Pd7×c6 must have been made long ago and his last move was either with K or R (1 Ke6) or with Pe5 from e7 (1 Pd5×e6 *en passant*). Again, NO!! If Black has played Pe7-e5, his pawn captures must have been f7×e6×d5×c4×b3×a2 and d7×c6, and White's six lost men include the dark-square bishop. So Black's last move was with K or R, and the solution is explicitly **1 Ke6**. "This took a host of scalps," wrote Dawson in 1943; mine duly followed many years later. Dunsany (1878–1957) was an Irish peer and a writer of highly evocative short stories, as well as a composer of chess problems.

Four Squares for Squares

Mogens Esrom Larsen

When the pharaohs built their pyramids with one stone at the top, four in the next layer, nine below them, etc., they stopped with 24 layers to make the total number of stones a square, 70^2.

As pointed out in 1919 by Dr. Watson, besides 1 this is the only square among these sums of squares.

Four Squares

Generalization to four dimensions leads to the question, if there are any squares among the sums of the first cubes? Already in 1631 Faulhaber knew the formula

$$\sum_{k=1}^{n} k^3 = \left(\frac{n(n+1)}{2} \right)^2$$

Four Cubes

A cube divided in each direction in four equal pieces is then parted in 64 small cubes—a square.

We may consider this number as the number of cubes in the two uttermost layers of the cube. Then we may ask the question, for which numbers, n, will the number of cubes in the two uttermost layers be a square, when the cube is divided in n^3 equal small cubes? As an example for $n = 10$ the number becomes $784 = 28^2$.

What is the function leading from a pair (n, m) satisfying the relation

$$n^3 - (n-4)^3 = m^2$$

Mogens Esrom Larsen writes the puzzle page "Une page d'énigmes" for the journal *Science Illustrée*.

69

to the next pair of solutions? We know the value of $(4, 8)$ to be $(10, 28)$.

Four Tesseracts

In dimension four the start of dividing the generalized cube—the tesseract-in four equal pieces in each direction, parting it into $4^4 = 256$ small tesseracts, gives the obvious solution 16^2. But are there any other solutions to the equation

$$n^4 - (n - 4)^4 = m^2?$$

Solutions

The map is

$$(n, m) \rightarrow \left(2n + \frac{m}{2} - 2, 6n + 2m - 12\right)$$

giving the solutions $(32, 104)$, $(114, 388)$, ...

Consider $p = n - 2$ to write the number of cubes as

$$(p + 2)^3 - (p - 2)^3 = 12p^2 + 16.$$

Then we ask for a square

$$3p^2 + 4 = r^2.$$

As any odd square satisfies $p^2 \equiv 1(8)$ this equation cannot have odd solutions. So we may write $p = 2s$ and $r = 2t$ to get

$$3s^2 + 1 = t^2.$$

An equation known as Pell's equation and solved by Fermat. The solution we already know is $(s, t) = (1, 2)$. Writing the equation as

$$(t + \sqrt{3}s)(t - \sqrt{3}s) = 1,$$

We find all solutions as

$$(2 + \sqrt{3})^k = t + \sqrt{3}s$$

with the recursion

$$(t + \sqrt{3}s)(2 + \sqrt{3}) = (2t + 3s) + \sqrt{3}(t + 2s).$$

The backtracking to $(n, m) = (2s + 2, 4t)$ gives the formula above.

That the equation

$$n^4 - (n-4)^4 = m^2$$

has no solutions with $n > 4$ was proved by Leibniz in 1678 in order to prove that the area of a primitive right triangle with integral sides is not a square.

Once upon a time I wrote to Martin Gardner asking if he ever used Pell's equation. His answer was "No!" This is my only two-way communication with him.

Bibliography

[Fau31] J. Faulhaber. *Academia Algebra.* Augsburg, 1631.

[Knu93a] D. E. Knuth. Johann Faulhaber and sums of powers. *Mathematics of Computation*, 61:277–294, 1993.

[Lar87] M. E. Larsen. Pell's equation: a tool for the puzzle-smith. *The Mathematical Gazette*, 71:261–265, 1987.

[Lei63] G. W. Leibniz. *Mathematische Schriften*, **7**, 261–265, 1863.

[Wat19] G. N. Watson. The problem of the square pyramid. *Messenger of Mathematics*, 48:1–22, 1919.

Part III

Smoked Ham: A Course in Magic

It's All about Astonishment

Tyler Barrett

What happens when you perform a magic trick for a puzzle solver? He tries to solve it! I discovered this when I began collecting puzzles. In the past 10 years I have traveled the world to find puzzles for my collection, and I have been fortunate to make many new friends with fellow puzzlers. Each year we get together to swap puzzles at what is known as the International Puzzle Party (IPP). These parties have been held annually for 20 years. For the evening entertainment at these parties, we usually have at least one magician perform. As an amateur magician for 35 years, I have been interested in my puzzle friends' response to magic. Almost universally they want to "solve" the tricks. This is unfortunate, for there is a key distinction between magic tricks and puzzles.

I believe that it was Professor Hoffman (Angelo John Lewis), well known to magicians and puzzlers alike, who tried to define the difference between a magic trick and a puzzle. To paraphrase: in puzzles, all the salient material is provided to the puzzler. In magic, some piece is concealed or held back from the audience. A puzzle is designed to be solved or figured out, while a magic trick is designed and presented with the goal of amazement or astonishment.

Since puzzlers have a solving perspective or bent, they usually view magic as a puzzle to be solved. The emotion that puzzlers are looking for is the "aha" or "eureka" when they make the necessary breakthrough to solve a puzzle. The emotion that magicians are shooting for through magic is amazement. Magic is performed to create moments of astonishment or amazement.

Tyler Barrett is president of Outside The Box Productions, a company that provides organizations with creative problem solving training which utilizes puzzles and magic.

In the December 1996 issue of *Genii, The International Conjurors' Magazine,* magician Paul Harris writes about this astonishment. He says that when a magic trick works, when you are truly blown away with a trick, there is a, "moment of ecstatic bliss where every thought (is) pulled from (your) face leaving nothing more than empty space." He likens these moments to childlike feelings we used to have before we grew up and "knew" so much. These moments are brief, and then our explanation machine kicks in and new thoughts crowd out the astonishment. It was only a trick. I know how he did that; it went up his sleeve. Her head didn't really come off her body; it was done with mirrors. All these "explanations" rush in to annihilate the astonishment. Our ability to categorize, pigeonhole and define cannot be put into abeyance for any appreciable length of time.

In their wonderful book, *Magic and Meaning,* Eugene Burger and Robert Neale explore the experience of mystery in magic. They point out that when magic is performed successfully, mystery is created. This mystery can be enjoyed in and of itself or it can be viewed as something to be solved. Burger writes, "Some audience members seem compelled to...ask themselves, their friends and even the magician how the trick was accomplished. Yet to experience this desire for the *how* is already to think of the mystery as a puzzle. Such people wish to solve the mystery, and so they confront the magic as a puzzle to be solved, and not as a mystery to be experienced." He continues, "My view is that a person in this...state *isn't* having an experience of magic at all. That person has stepped beyond the parameters of the magical experience...and has moved on to quite a different sort of experience: the analytical experience of attempting to figure something out."

Then there is the problem of verifiability. Let's say that you don't heed the advice above and are still compelled to "figure out" a trick. How will you know that you are correct? You can't ask a magician, because he holds to the Magician's Code that one never reveals a trick to a non-magician. You can suggest to your friends how you think it was done, but even if they agree, you still can't be sure you are correct. If you then hold to your explanation, you will have forever taken the astonishment, mystery and amazement out of this trick. I have witnessed this happen at a dinner table full of people, all convinced that they were in the know, yet completely wrong as to how the magic was accomplished. They had a solution, albeit a wrong one, and they had forever lost the magic.

So here's a puzzle for my puzzle friends. How can you stop trying to "solve" magic and just be astonished?

Bibliography

[BN95] Eugene Burger and Robert E. Neale. *Magic and Meaning*. Hermetic Press, Seattle, WA, 1995.

[Har96] Paul Harris. Astonishment is our natural state of mind. *Genii, The International Conjurors' Magazine*, 60(2):29–30, December 1996.

Shooting Craps Today

Russell T. Barnhart

When it opened on Saturday, October 19, 1996, as many as 75,000 gamblers streamed into the new Mohegan Sun Casino north of New London, Connecticut, and on the same day 50,000 streamed into its nearby rival, Foxwoods Casino. Who can say that casino gambling isn't popular in America today?

The four most popular tables games are craps, blackjack, roulette, and baccarat.

Though the history of crooked dice goes back to the cave man, let's begin with the invention of door pops by a con man named Finley in 1899. Finley made secret use of a midget whom he'd hide inside a cabinet in his hotel bathroom. Finley would then go downstairs to the hotel bar and confide to a select group of victims that he could throw a pair of dice any way they called them—and *after* the dice had left his hand! For instance, when the dice were in the air, if they called 8 with 6-2 or hardway with 4-4, that's exactly how the dice would land.

Naturally all the men at the bar scoffed and challenged him to prove it. So charging each of the scoffers ten dollars—a lot in 1899—Finley would lead them up to his hotel room, stand outside the closed bathroom door with a *borrowed* pair of dice in his hand, and as he threw the dice over the open transom into his bathroom, cried out, "Call 'em!"

Finley would always choose a bedroom whose bathroom door opened outward into his room. This way, when the suckers stampeded to open the door to see if they'd won their bets, they'd inevitably press him against the door, and he'd have to shout, "For God's sake, you guys, will you step back a way so I can open this door?" This delay would give the midget time to place the dice face up as 6-2, or whatever was

---

Russell T. Barnhart is a magician and mathematician who has authored several books on casino gambling strategies.

called, and climb back into the cabinet, locking its doors on the inside so nobody could find him. Then Finley would dramatically open the bathroom door, point to the dice on the floor, and declare, "See—6-2!"

Naturally the suckers would clamor for him to demonstrate it a second time—only once was probably just luck, a coincidence, they'd say. But feigning fatigue, Finley would reply that, as the feat required a great deal of concentration, he could perform it only once a night. So then they'd press him, and he'd finally give in and say that, just to prevent any unpleasantness, he'd be willing to accept another ten dollars from anyone to demonstrate it a second time.

So again Finley would throw the dice over the transom, someone would cry out a combination, the midget would position the dice for the correct total, and climb back in the cabinet. But finally people got wise. Rumor went round the downstairs bar that some sort of trickery was involved, that people were being hoodwinked, and Finley was no longer able to coax more suckers. Desperate, he figured he had to get rid of the midget and invent some less complicated method.

So that's why in 1899 Finley invented door pops—a pair of misspotted dice that, when thrown over a transom, will show only 7 or 11.

But how can that be?

As we know, every die is a cube with six sides, and the sum of the opposite sides of an honest die must always add up to 7. Thus the only combinations for opposing sides are 6-1, 5-2, and 4-3. But with door pops, one die has three 2's opposite three 6's, while the other has only 5's on all sides. Another combination is to put only 1's and 5's on one die and 6's on the other. Either way the dice can throw only 7 or 11.

So having fired his midget and moving his con game to an office building, Finley bet gamblers there that, after office hours, he could beat them on the comeout when, for fairness, he would throw a borrowed pair of dice over the transom. Then, on the floor of the ill-lighted office on the other side of the transom, he would point to the winning 7 or 11, casually pick up his door pops, and while handing them back to the sucker, switch them for the latter's honest dice and collect his bet.

How is it that such knowledge helps crap shooters today?

At any crap table a player can see at most three sides of a die. If at any time a player sees any two faces adding to 7, the dice are misspotted, so he should immediately get out of the game.

Are there any combinations that do the reverse, that don't throw 7 at all? Why yes. A popular misspotted pair carries on facing sides of one die only 3-4-5, and on the other only 1-5-6. This will through any total from 4 to 11 *except* 7 and craps, that is, except 7, 2, 3, or 12. But again whenever a crap player looks at this pair of dice, he'll see only two or three sides that add up to 7—so he should again get out of the game. A player must train himself to look not just at the tops of the dice but at the tops in combination with the sides too.

Having invented door pops, did Finley die a wealthy man? To the contrary, he died broke, as is the fate of most dice and card hustlers. While they're alive, hustlers love flashy clothes and cars and displaying a roll of big bills. They don't tell you that's all they have. Here are a couple stories about that.

A dice and card cheat meet on the sidewalk in a northeastern town. The dice hustler is from out of town. They stand and trade experiences of their recent triumphs. As they chat, a chauffeured limousine drives slowly past their curb. From the back seat a man smiles and waves at the card hustler. The car pulls away. "Who was that?" asks the dice hustler. "That's Raymond O. Smith. Enormously wealthy. Owns those woolen mills on the edge of town. He loses at least two grand a week in the poker game I run behind my pool room. He's a real sucker." The two hustlers trade more experiences. Another limousine drives past. From its back seat a man waves. The limousine pulls away. "Who was that?" asks the dice hustler. "That's John C. Brown. Owns the big bank two blocks down the street. He must lose at least a grand a week in the crap game at my place. A real sucker." Just then a bent, seedy, unshaven man shuffles past and pauses over a nearby trash basket. Rummaging in the basket until finding an abandoned hat filled with holes, he puts it on and drifts away. "Who was that?" "That's Poker-Faced Joe. He invented a way of dealing cards from the bottom of the deck invisibly. Smartest guy in town."

In the second story, years later the card and dice hustlers meet in the latter's home town. Again they trade anecdotes of their experiences. "What are your new angles?" asks the card hustler. "I've got a terrific new dice move. Been practicing it for twelve years—so naturally I've got it down perfectly. How about you?" "I've figured out a new way of marking cards. The marks are so tiny that even under a magnifying class you couldn't detect a thing. But what's your new dice move?" Come up to my room, and I'll show you." The dice man's room is nearby in a cheap rooming house filled with dust and dilapidated furniture. He points to a spot in the corner about a foot above the baseboards and puts a pair of dice onto the floor in front of his right foot: "What do you want? Name your number." "Ten," says the card hustler. "Five-five or six-four?" "Five-five." Taking careful aim, the dice hustler vigorously kicks the dice. They hit the right wall above the baseboard, careen into the left wall, and land on the floor: five-five. "Unbelievable!" cries

the card man. "Name your number." "Six." "Hardway or four-two?"
"Hardway." Again the dice hustler vigorously kicks the dice into the
corner baseboards. The two cubes careen off the right wall, bounce into
the left one, and land onto the floor: three-three. "That's the most
incredible work I've ever seen!" "Practiced twelve years," repeats the
dice man boastfully. "But then why are you living in a dump like this?"
The dice man shrugged tragically: "Can't find any suckers."

So much for being a crook. Good luck in an honest casino!

The Transcendental Knot

Ray Hyman

Introduction

For more than 30 years, one of my standard presentations that I give to college campuses and groups throughout the world is entitled *"Psychics" and Scientists*. The theme is how alleged psychics have fooled some first-rate scientists. I demonstrate typical phenomena that were described by these scientists. I then discuss the psychological reasons why such smart persons went badly astray. Because of time limits, I usually focus on one case as an example. This case is the one involving the German astrophysicist, Johann Carl Friedrich Zöllner who used his seances with the American spiritualist, Henry Slade, as evidence to support his theory of the fourth dimension. It was Martin Gardner who first inspired me to focus upon this case. The false knot routine that I describe here is one that I devised specifically to illustrate Zöllner's theory. In my routine, I follow carefully Zöllner's own explanation of his theory in terms of a perfectly flexible cord. Zöllner creates the analogy of a possible two-dimensional world—a flatland—to illustrate what can occur when another dimension becomes available. Zöllner's book was published six years before Abbott wrote his classic *Flatland* [Abb84].

Effect

The performer tells the story of Professor Johann Zöllner, a famous astronomer at the University of Leipzig in the 1870's. Zöllner developed

Ray Hyman is Professor Emeritus of Psychology at the University of Oregon and specializes in how smart people go wrong.

a theory of the fourth dimension, believing that many of the apparent miracles produced by spiritualistic mediums could be explained by the existence of a fourth spatial dimension. In his book, *Transcendental Physics* [Zol81], Zöllner tried to convey his concept of the fourth dimension in what he called "a theory of twisted cords."

Because humans cannot think in four dimension terms, Zöllner asked his readers to imagine beings who lived in a two-dimensional world. The performer displays a length of rope and asks the spectators to imagine a plane between himself and them. This plane represents the world inhabited by the two-dimensional beings. "Now, what would a knot look like in a two-dimensional world?" The performer rotates one end of the rope, always keeping it within the imaginary plane, in a clockwise manner until it crosses the middle of the rope and makes a loop. "This would be a knot in a two-dimensional world. The way to untie the knot in that world is simply to reverse the process—to take this end and make a counter-clockwise loop until the rope uncrosses itself. But, imagine that one of the two-dimensional beings suddenly develops the ability to gain access to a third dimension. He could untie this 'knot' by flipping the loop out of the plane and to the right so that the rope no longer crosses itself." The performer illustrates this. "To those individuals who are confined to their two-dimensional world, this sudden untying of the knot would appear to be a miracle."

"Zöllner then carried this analogy over to our three-dimensional world where we tie a knot by making a loop and then inserting one end of the rope through this loop." The performer illustrates by tying a simple overhand knot. "Now," Zöllner emphasized, "as long as we are confined to a three-dimensional world, there is no way to untie this knot except by reversing the process and putting the end back through the loop. As long as either end never goes through the loop, no matter how much we twist the knot it will still be there when we pull the ends apart."

The performer illustrates by twisting the overhand knot into the shape of a pretzel. Then he slowly pulls the ends apart and a knot remains, which he unties the normal way by inserting one end back through the loop.

"But," reasoned Zöllner, "what if some humans among us, such as spiritualistic mediums, developed the ability to make contact with a fourth spatial dimension? Then, they could use this fourth dimension to untie the knot in a manner that would appear impossible to those of us confined to a three-dimensional conception of our world." The performer illustrates his words by again tying an overhand knot, twisting the knot into a pretzel, and then slowly pulling the ends apart. The performer stops at a point when it appears that the simple overhand knot still remains in the center of the rope. He then shakes the rope gently and the knot disappears!

Requirements

A nylon rope, approximately five feet long. My rope is 3/8 inches thick, and I have wound thread around the ends to keep them from fraying. I have used the same piece of rope in over 200 performances since 1981 and it still remains in perfect condition.

Procedure

The two-dimensional "knot" is made by rotating one end in a clockwise manner until it crosses the middle of the rope and makes a loop (Figures 1–2). If the loop is simply picked up and moved to the right the "knot" will be "untied,"—that is, the rope will no longer cross itself. To make the real three-dimensional knot follow the procedure illustrated in Figures 1–4. To illustrate how twisting the knot cannot untie it as long as the ends remain outside the knot, start with the knot as in Figure 4 and with the right hand grab the rope at the points marked "R" in Figure 4 while the left hand holds the rope at the points marked "L." The position should now appear as in Figure 5. While holding the rope in this position, twist or rotate the top loops towards you for approximately three full turns. In other words, you use your thumbs and forefingers to twist the loop into a pretzel-like configuration. The use of a thick nylon rope enables the knot to hold this twisted condition. Figure 6 shows how this configuration will look to you after you regrip the ends of the rope in your hands and allow the twisted knot configuration to fall forward and hang between the ends in the middle of the rope. If you now move your hands to the ends of the rope and let the rest hang down, the resulting configuration should look to you something like Figure 6. Gradually pull the ends apart until the "pretzel" becomes untwisted leaving the original overhand knot in the middle of the rope. To untie such a knot in a three-dimensional world one must put the right hand end of the rope through the loop and pull the ends apart to untie the knot. Openly thread the right hand end through the loop and show that the knot is untied.

Now, following the story line, apparently tie the overhand knot once more. But, this time you follow Figure 1 with Figure 7. Once the first

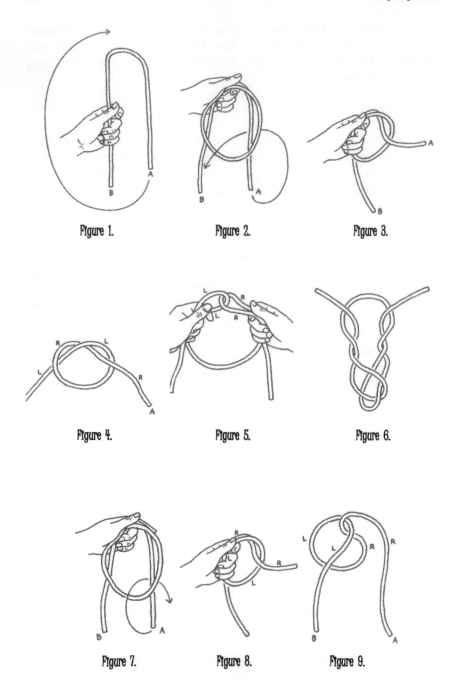

Figure 1.   Figure 2.   Figure 3.

Figure 4.   Figure 5.   Figure 6.

Figure 7.   Figure 8.   Figure 9.

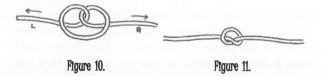

Figure 10.   Figure 11.

loop is made, insert end "A" through the loop from the back to the front (from your side towards the front). This looks like a simple overhand knot, but is actually a simple loop that will vanish if the ends are pulled apart. The situation is now as pictured in Figure 7, which looks very much like a true overhand knot. You can even pull end "A" to tighten the knot a little. Do not make the loop too small, because you want it to be sufficiently large for the pretzel business. Follow the same moves depicted in Figures 4–6 to make the twisted knot or pretzel. (Figures 7 and 8 correspond to Figures 4 and 5 which were made with the true knot.) Now, slowly pull ends "L" and "R" apart to enable the pretzel to gradually untwist. You will find that you can stop the process at two or three stages. One stage is shown in Figure 9. It looks like a real knot still is tied in the middle of the rope. After a few trials you should find that you can actually stop at the stage depicted in Figure 10 which looks much like a true overhand knot. (Figure 11.) It is at this stage that you pause as you explain that, for us three-dimensional beings, a knot still has to remain. But, if someone actually had access to a fourth spatial dimension, he could simply flip the knot into that dimension, and it would become undone. The performer suits his actions to the words and gives the rope a slight shake and knot disappears.

All this may sound complicated, but if you get the right size nylon rope and follow the instructions, you will see how effective this can be. You will often not be sure yourself if the knot in the rope is real until you shake it out.

Comments and Credits

I devised this effect as part of a longer routine that I do about Professor Zöllner. For more information about Zöllner, as well as additional ideas for effects built around the theme of the fourth dimension see my article, "The Zöllner Phenomenon" [Hym85]. In addition to a routine for producing knots in a closed loop or rope, an effect which the medium Henry Slade used to convince Professor Zöllner that he had access to the fourth dimension, I supply a bibliography of nine useful sources.

I got the idea for this routine from Zöllner's own book *Transcendental Physics*. Ron Friedland, after witnessing my first awkward version, helped me devise the phase of the routine that begins with Figure 8. At that time, however, we had no way to arrive at this position without first arranging the rope into the configuration upon a table. A few days later, while working with the rope, the obvious solution, the one described here, came to me all of a sudden. That was on September 14, 1981, and I have been doing it the same way ever since.

In preparing this Parade, I recently did some cursory research in my library to find predecessors for this version of a vanishing knot, I found close versions of the false overhand knot as depicted in Figures 6 and 7 in Knot that is Not [Ric41], The Phantom Knot [Tar44], Grant's Ghost Knot [Tar54], and The Illusive Knot [Jamte]. I was sure that the twisting, pretzel aspect of the routine was entirely original with me. But to my surprise, I found something similar in Joe Cossari's Naughty Knot [Jam80].

This is a slightly edited version of my effect The Pretzel Knot which first appeared, along with a number of other of my creations, in *The Linking Ring*, October, 1986 [Hym86]. *The Linking Ring* is the official publication of the International Brotherhood of Magicians. The drawings were made by Marshall Philyaw.

Bibliography

[Abb84] Edwin Abbott Abbott. *Flatland; a romance of many dimensions.* Dover, 1992 (1884). Reprint.

[Hym85] Ray Hyman. The Zöllner phenomenon. In Stephen Minch, editor, *The New York Magic Symposium: The Collection Four*, pages 92–96. Richard Kaufman and Alan Greenberg, New York, 1985.

[Hym86] Ray Hyman. The pretzel knot. *The Linking Ring*, 66(10):80–84, 1986.

[Jam80] Stewart James. *Abbott's Encyclopedia of Rope Tricks*, Volume 3. Abbott's Magic Novelty Company, Colon, Michigan, 1980.

[Jamte] Stewart James. *Abbott's Encyclopedia of Rope Tricks*, Volume 2. Abbott's Magic Novelty Company, Colon, Michigan, No date.

[Ric41]   Harold R. Rice. *Rice's More Naughty Silks*. Silk King Studios, Cincinnati, 4th edition, 1941.

[Tar44]   Harlan Tarbell. *The Tarbell Course in Magic*, Volume II. Louis Tannen, New York, NY, 1944.

[Tar54]   Harlan Tarbell. *The Tarbell Course in Magic*, Volume VI. Louis Tannen, New York, NY, 1954.

[Zol81]   Johann Zöllner. *Transcendental Physics*. New York: Arno Press, 1976 (1881). Reprint, translated from German.

The Brazilian Knot Trick

Maria Elisa Sarraf Borelli and Louis H. Kauffman

Introduction

In this paper we introduce a significant variant on the famous Chefalo knot trick [Ash44] that we dub the Brazilian knot trick. It happened this way: The first author of this paper (a Brazilian of Italian descent), whom we shall refer to as MEB was being shown the Chefalo knot by second author, whom we shall refer to as LK. All this happened long ago and far away in the year 1996 at an International Conference on Mathematical Physics and Knot Tricks held at Cargese, on the island of Corsica, in the midst of a startlingly blue sea. But we digress. LK, having demonstrated the disappearing nature of the Chefalo knot, said to MEB, "Now you do it!" MEB took up the rope and tied a knot that we saw at once was not the Chefalo knot. But even though it was not, it was in fact not a knot and this new knot that was not a knot became the Brazilian knot—the subject and object of our paper.

Brazil Knots

We begin with Figure 1, a depiction of the original Brazilian knot as tied by MEB on the lunch table in Corsica, under the noonday sun by the deep blue sea. It surprised us a bit to see this knot appear. MEB thought it quite natural at the time. LK was perturbed, but determined to push on. And so he said, "Well maybe your Brazilian version can still be threaded to make an unknot that will amaze us!"

Louis H. Kauffman is a leading knot theorist who has authored the two books *On Knots* and *Knots and Physics*," and edits the *Series on Knots and Everything*. **Maria Elisa Sarraf Borelli** has just finished her Ph.D. in Physics.

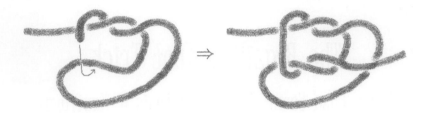

Figure 1.   Brazilian knot.

And so they looked. And it did make an unknot by a little further threading just as shown in Figure 2.

Figure 2.   Brazilian unknot.

But now, says LK, we have to add the ring, and find a threading that will be unknotted *and* release the ring. "Must we?' says MEB. "We must," says he. See Figure 3.

Figure 3.   Brazil with a ring.

And eventually they did. It took some time. And the stars showed over the deep blue sea. The ringed release knot is shown in Figure 4. Try it and you will see.

Figure 4.   The Brazilian unlinking ring.

A Few More Twists

There are infinitely many Brazilian knots and unlinking rings! In the construction of the Brazilian knot we made two windings, each a full turn, on the upper and lower portions of a bend in the rope. The second winding can be done any number of times and it can also be done in reverse. This means that there is a Brazilian knot and knot trick for each integer number N where N denotes the number of times one performs the second winding. The original Brazilian knot and unknot are the case of $N = 1$.

The reader who looks closely at our pictures of the various cases will note that a winding labelled N full twists consists in $2N + 1$ half twists. Normally, $2N$ half twists make N full twists, but in the handling of the rope one begins the winding on one side of a strand, and ends it on the other side (imagining the rope laid nearly flat on a table). This movement from side to side creates the extra half twist in the winding.

Figure 5 shows the case that we call $N = -1$, a Brazilian analogue of the Granny knot. In this Figure we show both the knot and the unknot for the case of $N = -1$. In Figure 6 the ring release is shown for this case. It is interesting to note that the Granny knot version of the original Chefalo trick does not in any obvious way lead to an unknot or to a ring release. See the Appendix for a discussion of this point. In Figure 7 we show the case of the ring release for $N = 2$ and indicate the general case of arbitrary N. Figure 8 shows the Brazilian Unknot in the case $N = 2$. In Figure 9 we show the setup for the ring release with $N = 2$. We trust that these examples and a little ropework will convince the interested reader that there is a Brazilian knot trick for every integer (including zero)! A combination of $N = 1$, $N = -1$ and $N = 2$ makes a spectacular demonstration.

Figure 5. $N = -1$: The Brazilian Granny knot.

Figure 6.   Ring release for $N = -1$.

Figure 7.   General N and $N = 2$.

Figure 8.   The $N = 2$ Brazilian unknot.

Figure 9. The ring release for $N = 2$.

A Note on Performance

The knots and unknots herein described can be performed by tying them on a length of rope whose ends are then held fast by an assistant to the magician or by the magician herself. An unknot is demonstrated to be unknotted by pulling the ends of the rope until the "knot" in the middle disappears. A ring is demonstrated to be unlinked with the rope by pulling the ends, while an assistant helps the mass of rope in the middle disentangle by a little gentle encouragement, until the ring drops off and the rope is seen to be unknotted. Another method in all cases is to tie the ends of the rope together securely. The unknot or ring-release unknot can then be manipulated at will by the magician or spectator until it reveals its secret!

Appendix on the Chefalo Knot Trick

The original Chefalo knot trick is performed by starting with a version of the square knot as shown in Figure 10. The corresponding unknot is also shown in Figure 10. In Figure 11 we show the Granny knot as a starting position. There is no direct analogue of the unknot and ring release known to us that starts from the traditional Granny. The Brazilian Granny as shown in Figure 5 is our case $N = -1$ and works fine. The Brazilian Granny does the trick.

Acknowledgements

Louis Kauffman thanks the National Science Foundation for support of this research under grant number DMS-9205277 and the Institute Henri

Figure 10.   The Chefalo unknot.

Figure 11.   The traditional Granny.

Poincare and Institute Emile Borel for providing the superb facilities under which this paper was written. Maria Elisa Sarraf Borelli and Louis H. Kauffman thank the Nato Advanced Study Institute held at the Institut d'Etudes Scientifiques de Cargese from September 1 to 14, 1996 for generous support. We thank Pierre Cartier, Cecile DeWitt-Morette, Hagen Kleinert, John H. Conway, Naomi Caspe, Doug Kipping, Claude Bourgeois, and Diane Slaviero for invitations, introductions, square dancing (see the article by LK in [Kau97]), rope, and helpful conversations. Both authors thank their lucky stars.

Bibliography

[Ash44] C. W. Ashley. *The Ashley Book of Knots*. Doubleday, New York, London, Sydney, Auckland, 1944.

[Kau97] Louis H. Kauffman. Functional integration—basics and applications. In Pierre Cartier Cecile DeWitt-Morette and Antoine Folacci, editors, *NATO ASI Series—Proceedings of the Nato Advanced Study Institute held at the Institut d'Etudes Scientifiques de Cargese from September 1 to 14, 1996*, Plenum Press, New York and London, 1997.

Mental Match-up

Meir Yedid

This is an effect that I held in reserve for more than ten years. It is an excellent and almost impromptu mental mystery. It uses only a few simple props but involves three spectators in various tasks and plays very big. I often uses it in platform or stage shows.

Because the routine will move along different lines depending upon the outcome of actions undertaken by members of the audience, I will not try to outline the effect in detail. Read through the presentation and you will quickly grasp the underlying concept. Once this is understood, I am sure you will want to give the routine a try. Once you try it, I am sure you will perform it often.

Props and Preparation

You will need the following items: a standard ESP deck, a jumbo ESP card (which you can either make or buy), a large Manila envelope (6 inches by 9 inches), a short magic marker or pen, and at least one complete book of matches.

The symbol on the jumbo ESP card will be the card that is forced from the ESP deck. Let's assume it's the circle. Arrange the ESP deck so that the circle appears in the first five even positions from the top of the face-down deck (i.e., positions 2, 4, 6, 8, 10). In addition, make sure that the cards directly above and below each of the five force cards are different from each other. A sample arrangement of the ESP deck might be:

Meir Yedid is a magician and author who amazes audiences with his fanciful fingers ... and lack thereof. This article was reprinted from Meir Yedid's *Magical Wishes*, written by Stephen Hobbs, available at http://www.mymagic.com/yedbooks.htm.

star → circle → square → circle → wavy lines→ circle →
cross → circle → square → circle → rest of the cards (in any
order)

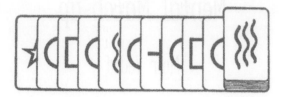

Place the ESP deck in its case.

On the outside of the large Manila envelope write, "PREDICTION."
Open the envelope, stick your hand inside (which is why it must be
large), and write the following prediction on the inside of the envelope:
"THERE ARE NINE MATCHES LEFT IN THE MATCHBOOK." Place
the jumbo ESP card inside the envelope and seal it shut.

Finally, make sure that you have at least one complete book of
matches at hand.

Performance

I will describe this as I present it in a platform or stand-up show. The
small procedural changes required when performing in more intimate
environments should be obvious.

Give the prediction envelope to an audience member. Have her con-
firm that it is sealed shut and ask her to hold it in plain view until the
conclusion of the experiment.

Bring out the ESP deck. It is important that the audience be famil-
iarized with the ESP cards. I usually talk about Dr. Rhine's research
into the paranormal and how simple symbols were used in the design of
the deck on the theory that they would be the easiest objects to trans-
mit telepathically. Make sure that the audience sees the various symbols,
but do not tell them that the deck consists of only five symbols repeated
five times. As you finish speaking, casually overhand shuffle the face-up
deck without disturbing the ten-card stock at the rear (top) of the deck.
Place the cards into their case and hand the case to another spectator
with the request that she hold it for a few moments.

Ask if anyone in the audience has a book of matches. I always bring
out my book of matches at this point. This visually confirms your request in case anyone is confused and thinks that you want a box of
matches or a lighter.

Let's assume for the moment that someone does have a book of matches. You want to be well away from this person for the next few minutes, so if he or she is in the front row move away from them as you say, "Would you please open the matchbook and tell me whether it is unused, or whether some matches are missing from it."

If the spectator says that there are some matches missing you respond as follows, "Great. There is no way that I can know how many matches are left in the matchbook."

If the spectator says that the matchbook is complete say, "Will you please tear out some matches and throw them away."

Now let's jump back for a moment. If no one in the audience has any matchbooks, then simply toss out your own book of matches. You know this book is complete, so ask a spectator to tear out a few matches and toss the matchbook to another spectator. Emphasize that you could not possibly know the number of matches remaining in the book.

Regardless of the procedure followed, the spectator is now holding a matchbook that contains between one and nineteen matches. Continue by saying, "Please count the matches remaining in the matchbook silently to yourself and remember how many there are."

When the spectator has counted the matches say, "So you are thinking of a number of matches. I assume it's a two-digit number?" This question is asked casually, but the spectator's response is critical.

If the spectator says that it is not a two-digit number, then you will not be able to use the second prediction—on what is hidden inside the envelope. Act surprised and say, "Well, okay. Would you please tell everyone the number you are thinking of and which only you could possibly know."

If the spectator confirms that it is a two-digit number, then you will be able to use the hidden prediction but must also go through an additional bit of procedure. Say, "I want you to add the first and second digits of the number together. So, if you have twenty-three matches then you would add two and three together and get five. Okay?" The example "twenty-three" is a throw-off as the spectator will never have that many matches. When the spectator has completed the calculation say, "Now tear out a number of matches from the book equaling this new number and hold them tightly in your hand." The spectator will tear out a number of matches ranging from one to ten (you do not know how many) and there will always be nine matches remaining in the book.

Recap what has taken place so far—prediction, ESP cards, random number of matches—and then say, "For the first time, will you tell us how many matches you are holding." The spectator reveals the number.

Turn to the spectator holding the ESP deck (you may want to invite her on stage) and ask her to remove the cards from their case. Have her count the cards onto the table one at a time, stopping her when she reaches the number named by the spectator. You now turn over one of the force cards, which will either be the last card dealt or the card remaining on top of the deck. Since you know that the force cards occupy the even positions, this is a simply matter of keeping track of the cards as they are dealt. "You stopped on the circle." Turn over the cards directly above and below the force card. "If you had gone one card more or less you would have stopped on a _____ or a _____." Here you name the two cards that are adjacent to the force card which, thanks to the set-up, will always be different from the force card and from each other.

Ask the spectator to open the prediction envelope, take out what is inside, and display it. She removes the jumbo duplicate of the force card. If the spectator had less than ten matches in the matchbook, then this is the end of the routine so play it big.

If, however, the spectator had ten or more matches in the matchbook, say, "An amazing coincidence? Perhaps, but perhaps not. But I made a further prediction which is also inside the envelope. Would you please take that prediction out." The spectator looks inside the envelope but, because she is expecting another card, does not see anything. "No inside, on the envelope itself. Rip it open." She rips open the envelope. Have her read the prediction out loud.

Turn to the spectator with the matchbook, "Please count the matches remaining in the book out loud one at a time." He does so, and the slow progression towards nine serves as an ideal applause cue.

History and Inspiration

Those familiar with the "nine principle" will recognize its use here, although it is nicely disguised. The nine force works as follows: For any number between 10 and 19, the sum of the two digits subtracted from the number itself will always equal nine. Moreover, the sum of the digits is always a number from one to ten. The nine principle is traditionally used to force the number nine. I believe I am the first person to use "both ends" of the principle—i.e., to also use the fact that the sum of the digits is never more than ten.

It just so happens that most matchbooks contain exactly twenty matches. (Make sure that the matchbook you use is not one of the larger or smaller ones used as promotions.) This means that the nine force can always be used if at least one match is missing from the book. Also, since the ESP deck contains five duplicates of each symbol, positioning the force symbol in the first five even positions means that the last force card is the tenth card from the top of the deck. If it were any further, it could not be forced using the number derived from the nine principle. The matchbook and ESP deck just happen to work well together.

Finally, it may seem as if the double ending will be achieved less than half of the time. In practice, it is reached much more often. The kicker ending will be achieved if their are ten or more matches in the book, and the routine takes advantage of that fact in several ways. First, if you are handed an incomplete matchbook, it usually has more than ten matches in it. I've noticed that people tend to throw away a matchbook if it has only a few matches remaining. Second, if the matchbook is full and you ask the spectator to tear out some matches, the spectator—in the interests of expediency—usually tears out only a few matches, again leaving you with more than ten matches in the book.

The idea of using the nine principle with a matchbook was first published by Fred DeMuth in *The Jinx* (August, 1935), although I didn't know that when I constructed this routine.

A Labyrinth in a Labyrinth

Gordon Bean

Effect

As a spectator shuffles a deck of cards, the performer begins: "If one were lost in a labyrinth, a deck of cards would be a handy thing to have. You could leave a trail of cards behind you—like bread crumbs—to help you find your way out. Interestingly, a shuffled deck of cards is itself a labyrinth. Every shuffled deck has a unique order. It is hard to make your way from beginning to end of a randomly shuffled deck with no false turns."

The performer takes back the deck and openly peruses batches of cards. "But if you step back and see the big picture, it's often surprisingly easy to escape from anything—even a deck of cards." The performer proceeds to deal the cards face up on the table in a twisting trail.

"Like many journeys, this one is determined by your first step. Please point to one of the cards in this first section. We're then going to spell the value of that card—say t-w-o. We will then spell the value of the card we land on, and continue in that manner, going through the entire trail, until we get to a card that doesn't lead to another card. That will be the end of the labyrinth."

The spectator points to a card, the performer offers an opportunity for the choice to be changed, then together they spell their way through the deck as described above until they reach a card which does not

Gordon Bean lives in Los Angeles, where he is the librarian at the Magic Castle. He also runs Bean's Magic, which markets his magical inventions. This article first appeared in *Labyrinth: A Journal of Close-Up Magic* [Bea99]; the original text by Stephen Hobbs has been altered here by Gordon Bean.

spell to another card: the seven of spades. "That's it," the performer announces, "The end of the labyrinth. Of course, it's easy for me to say. I got to see the big picture." The performer invites the spectator to come around and see the view from the other side of the table. The, labyrinthine trail of cards is seen to consist of a big "7" joined to a big "S."

Performance

This is based on the Kruskal Principle. This is the tendency, discovered by Dr. Martin D. Kruskal, for various chains in a shuffled deck to intersect, making them likely to end on the same card. This performance includes two new notions. First, the presentation gets away from both mathematics and dealing through the deck. Second, the chance of success is increased from a solid probability to an almost dead certainty. See the Notes & Credits for more information in this regard.

To perform, take back the shuffled deck and begin spreading the cards in batches from hand to hand. As you do so, casually shift the seven of spades to the face of the deck. This is the only move in the routine. (You can also shift the seven of clubs, changing the layout accordingly.) The motivation for looking at the faces of the cards is your statement that you are trying to see the big picture contained within the labyrinth of the shuffled deck.

Begin to deal the cards, lengthwise, into the pattern depicted in Figure 1. As you deal the cards, start silently spelling the values (and only the values) of the cards, beginning with the third card. Thus, if the third card were a four, you would start spelling "f-o-u-r". The "f" would be the fourth card, the "o" the fifth card, and the "u" the sixth card and the "r" the seventh. If the seventh card was a nine, you would continue the spell with "n" on the eighth card, and so on.

Continue in this manner, silently spelling the values of the cards, until you have five or fewer cards left in your hand. Flip the remaining cards in your hand face up and deal the seven of spades as the last card of your silent count. If you get really lucky, this might even be the last card in the deck. Deal out the remaining cards, completing the pattern.

That's it—the effect is now performed as described above. The spectator picks any card in the upper arm of the seven as a starting point. Begin the spelling procedure, out loud, from that point. Things will go more smoothly if you do the spelling, but make sure your actions are fair and clean. Eventually, and usually fairly quickly, the out-loud spell will mesh with your previous silent spell and you are home free. The

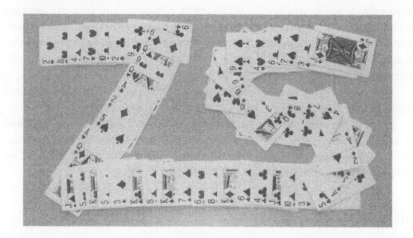

Figure 1.

out-loud spell will end up on the seven of spades. You can't spell any further and have completed the maze. The spectator is then invited to your side of the table to see the "big picture" contained in the labyrinth itself.

Very occasionally, the verbal and silent chains won't mesh and the effect won't work. If this happens, simply have the spectator pick a new starting card, saying: "Do you see how difficult it is to get out of a maze? But if you step back and take a look at the big picture and then pick a new opening card, you should have no trouble at all."

Notes and Credits

The Kruskal Principle was originally described by Martin Gardner in the June 1975 issue of *The Pallbearers Review* [Gar75], with additional comments by Karl Fulves (this article was later reprinted in *Martin Gardner Presents* [Gar93], sans Fulves' comments). See also Martin Gardner's February 1978 column in *Scientific American* [Gar78a].

The Kruskal Principle has traditionally been performed by counting the values of the cards. By changing this to spelling the values of the cards, this presentation radically improves the chance of reaching a successful conclusion. When you count numerically, you are working with thirteen unique values (although in the original, all face cards were assigned a value of ten, which Fulves suggested be lowered to five). In

comparison, when you spell—although it appears that you are working
with thirteen possibilities—there are actually only three different possibilities (three letters, four letters and five letters). This reduced field of
possibilities means that it is much more likely that different strings will
intersect.

The ever fertile mind of Martin Gardner has applied the Kruskal
Principle to spelling words in a paragraph of text [Gar98].

Finally, the idea of having a design hidden in the layout—as opposed
to a design being progressively revealed—is one that I hadn't encountered before. Max Maven, however, pointed out that a related idea by
Tony Koynini appeared in the *Magic Wand* [Koy53].

Bibliography

[Bea99] Gordon Bean. A labyrinth in a labyrinth. *Labyrinth: The Journal of Close-Up Magic*, 11:4–6, 1999. Text by Stephen Hobbs.

[Gar75] Martin Gardner. The Kruskal principle. *The Pallbearer's Review*, 10(8):967–970, 1975.

[Gar78a] Martin Gardner. Mathematical games. *Scientific American*,
238(2):19–32, February 1978.

[Gar93] Martin Gardner. *Martin Gardner Presents*. Kaufman and
Greenberg, Silver Spring, MD, 1993.

[Gar98] Martin Gardner. A quarter-century of recreational mathematics. *Scientific American*, 279(2):68–75, August 1998.

[Koy53] Tony Koynini. Koynini's karpet. *Magic Wand*, 42(237):42,
March 1953.

Paradox Squares Force

Robin Robertson

Paradox Papers

Martin Gardner's "Paradox Papers" was first published in the July 1971 issue of the magic magazine *The Pallbearer's Review*. It presented a new topological principle in magic using nothing more than a square sheet of paper [Gar71]. Gardner added some additional ideas and variations in [Gar83, pp. 71–73]. I'll summarize the principle in its most elementary form, then present a way to force a number using this principle.

Start with a square piece of paper. A good size is to take a normal 8-1/2×11 inch sheet and cut it to an 8-1/2 inch square. Fold it in half twice in each direction, forming 16 squares. Make all the folds several times in each direction, so they fold easily along any crease.

Mark large X's and O's in alternating squares like this:

Then turn the paper over—it doesn't matter if you turn it over sideways, or end-over-end—and mark X's and O's on this side the same way. You'll find that an X on this side is backed by an O on the other side, and vice versa.

Robin Robertson is a psychologist, magician, mathematician, and writer who has published ten books and over one hundred articles.

Now fold the square along the lines **IN ANY WAY** you like to make a packet that is only one square high and wide. Just as one of many possibilities, you might fold the left column over to the right, making a figure 3 columns wide and 4 columns high. Then fold the top row down under, and so on. Any way will do. When you're finished, take a pair of scissors and cut all four edges. Make sure you cut away all folds, as it's easy to miss some if you're not careful.

If you examine the little squares of paper, you'll find that all the X's are face-up and all the O's face-down (or vice versa). That's the basic principle and Gardner explained several possibilities using the principle. Here's another.

Paradox Squares Force

Assume that instead of X's and O's, you're going to fill the 32 possible squares (i.e, 16 on each side) with unique numbers from 1 to 32. Their sum then will be $(32 + 1) \cdot 32/2 = 33 \cdot 16 = 528$. But, more importantly, if you fill in the sides carefully, when folded and cut, either side will total exactly half this, or 264.

In order to do that, simply make sure that if any number N goes in an "X" square, $33 - N$ also goes in an "X" square, on either side. Similarly, if M goes in an "O" square, $33 - M$ also goes in another "O" square. Here's an example of a filled-in square:

SIDE A			
11	31	8	21
5	27	18	20
1	14	4	28
9	25	16	13

SIDE B			
32	26	6	17
12	10	2	23
3	7	22	24
19	30	15	29

Presented as a magic trick, you write a prediction of "264" in advance, seal it in an envelope and ask someone to hold it. Then bring out the square and allow another person to fold it any way they like, then cut off the edges. Have someone read off the numbers on top of the 16 squares while another person adds them using a calculator. They'll come up with "264". You then have the envelope opened and the prediction read. Sure enough, you predicted "264" even though you had no way of knowing how the paper would be folded.

Bibliography

[Gar71] Martin Gardner. Paradox papers. *The Pallbearer's Review*, July 1971.

[Gar83] Martin Gardner. *Wheels, Life and Other Mathematical Amusements.* W. H. Freeman, New York, 1983.

A Face in the Shadows

Larry White

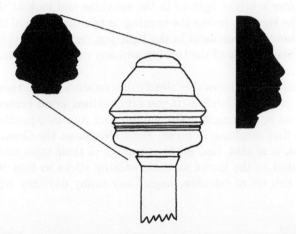

A while ago my eyes happened to fall upon a wonderful "profile turning" my late Dad made me many years ago. This was a four-inch-high piece of maple which he had turned on his wood lath. It looked like a small solid vase or urn of some sort and it became an object of curiosity for many future visitors to my home. They would pick it up, turn it all around, study it—puzzled because it seemed to have no utilitarian use whatsoever—and always had to ask what it was. "It has a secret," I would explain, "And once you know the secret you know what it is." Then I would show them the secret and, although no magic gimmick was involved—the secret, of course, was the shadow—it always produced the

Larry White is the Magic Editor of *M-U-M*, the official magazine of the Society of American Magicians, and the Science Editor for *HOPSCOTCH* and *Boy's Quest* magazines. In a slightly different form this article was published in the February 2000 issue of *M-U-M*, the official magazine of the Society of American Magicians.

same ooohs and ahhs any good magic effect garners. It became Abraham Lincoln!

If you see a profile of a well-known person you can immediately identify that person. But suppose the "profile" is visible from all sides, 360 degrees? Your brain no longer recognizes it as a profile. So, if a craftsman cuts a template of a person's profile and uses this template to create a wood turning each detail will be scored completely around the wood and this is called a "profile turning," though the profile will not be recognized without some help.

There are two ways to see a recognizable profile in a profile turning. You can mask half of the turning with your hand and squint at the remaining half. With practice you will see the profile if you look *only* at the edge. The far easier and more startling way is by holding the turning before a bright light or in the sunshine and look at the shadow it casts. The shadow, unlike the turning, is two-dimensional rather than three. By holding your hand in the light you can cast a shadow to blot out 1/2 of the shadow of the turning and any onlooker will *instantly* see the profile.

The illustration shows this clearly. To retain a magic theme I chose to use the profile of Houdini. If you are a skilled wood turner, or know someone who is, you might make one showing your own profile. You will not be the first magician to do so. Both Herrmann the Great and John Mulholland, it is said, had profile turnings of their faces made. These were mounted as the knobs on their walking sticks so they were ready to present this bit of "shadow magic" any sunny day they went out for a stroll.

A Card Vanishing in a Nut

Bob Friedhoffer

The accompanying routine is probably the first practical joke related to a magic trick to be found in an English magic book. The book in which it is found is the among first two or three books in the English language to deal solely with magic.

There has always been a bit of conjecture on the name of the book. It is commonly known as *Hocus Pocus Junior* to magicians, though the title page makes it appear that Hocus Pocus Junior is the pen name of the author, with the title being *The Anatomy of Legerdemain or the Art of Juggling*.

The pages that follow are from the eighth edition printed in London, which, according to Raymond Toole Stott, was printed circa 1671. The only known copy of this edition was found in the New York Public Library.

I've taken the liberty of changing the "f" to the modern "s". Spelling, grammar, and punctuation appear as in the original.

The reader should be aware that in the 1600s nuts were usually opened by the masses by placing a nut in the mouth and cracking it with the teeth.

How to make a card vanish, and find it again in a Nut.

Take what card you will, pill the printed paper from off of it, and role it hard up, and make a hole in a Nut, and take out the kernel, and then thrust in the Card, afterwards stop the hole of the nut neatly with wax, this nut you must have

Bob Friedhoffer is a leading proponent in the use of magic as a tool to teach creative thinking techniques and scientific principles to the public.

in readiness about you, and when you are in your play, call
for such a Card as inclosed in your Nut, or else have on in
readiness, and say, You see Gentlemen here is such a card:
then wet it and pill off the printed side, role it up, and in
the usual manner convey it away: Then take your Nut out
of your pocket, and give it unto one, and say, crack that Nut
and tell me if you can find the Card there, which being found
will be found very strange.

Then have another such like Nut, but filled with Ink, and
stopped after the same manner that your other Nut was, and
give that unto another, and bid him crack it, and see what
he can find in that, and so soon as he hath cracked it, all the
ink will run about his mouth, which will move more mirth
and laughter then the former.

Casey at the Fox

Ken Fletcher

Background

At the Gathering for Gardner two years ago, I was browsing through some books authored by Martin Gardner and saw a book titled *The Annotated Casey at the Bat,* and wondered how a book about my favorite sport had mistakenly slipped into this group of books about magic and mathemagic. Regardless, I knew I wanted it and picked it up. I can't begin to tell you how amazed I was when I saw the book was edited by Martin Gardner! I read every ballad and every parody of the Mighty Casey as Captain Ahab, a football player, a cricket player and even a cosmonaut. I thought a magician, The Great Casey, would be a *natural* (pardon the pun) to add to this team of parodies.

Over a quarter of a century ago I called Martin "out of the blue" for help on a mental effect I had created and was planning to market. Without even knowing me he invited me up and gave me some great tips. In this very specialized field there are people that help and encourage newcomers and ask for nothing in return. Martin Gardner is one of those people and has helped so many at every level. Thank you for your help and encouragement so many years ago and thanks for the book that inspired The Great Casey. It was phun.

Ken Fletcher is the founder and President of Magic Masters, Inc. and creator of the famous Rocky Raccoon.

Casey at the Fox

The crowd was pretty hostile to the Vaudeville nine that day;
The M.C. blew every gag—they needed Max or Dan or Jay.
And when the opener had bombed and the second act had
    failed,
A disappointed grumbling throughout the house prevailed.

A struggling few got up to go in deep despair.
The rest remained to see it through and said a silent prayer.
"Where's the Great Casey? He'd quell these angry hoards.
We know we'll get our money's worth if Casey strides the
    boards."

But Mimi the Mime preceded Casey as did the Juggler Jake,
And the former was a yawner and the latter was a flake.
Every soul within the Fox thought their prayers had gone for
    naught;
For there seemed but little chance they'd see Casey in the
    spot.

But Mimi knocked 'em dead as she walked into the wind,
And Jake caught every ball and club and each face now wore
    a grin.
And when the clapping ended and the curtain had come
    down,
They knew they'd soon see Casey and his wonders world
    renown.

Then from 500 throats and more a lusty cheer grew in the
    Fox.
It rumbled in the balcony, in the loge and every box;
It knocked upon the rafters and recoiled from each footlight,
For Casey, The Great Casey, had entered from stage right.

There was ease in Casey's manner as he strode into the light;.
There was pride in Casey's bearing and his smile was shining
    bright.
And when a flash appeared and a rose plucked from the glow,
No one in the crowd could doubt Casey would save the show.

Then with a haughty flourish both empty hands he showed
And held aloft a silk top hat and to center stage he strode.
But the rabbit had quit the act and left a souvenir impure,
And Casey's hand went deep inside and spoiled his manicure.

A few up front saw Casey scowl in this situation sticky.
I'll split that hare, thought Casey, and I'm not just being
     picky.
But he smiled and set the hat aside...they'll think I've yet
     begun;
But in the wings they saw the goof and someone said, "That's
     one."

Two silver rings are now displayed, one in each of Casey's
     hands,
And from the darkened room there came a cheer from Casey's
     fans.
Each one had seen the rings before—to them 'twas history,
But all still wondered at the feat and enjoyed the mystery.

Casey brings the silver bands together till they meet;
One ring inside the other—it is a wondrous feat.
Each mouth hung wide in silence as the rings just met and
     clinked.
Five hundred minds within the Fox knew they should have
     linked.

Now Casey uses all his skills to hold back profanity;
For he knows the rings he held were just a bit "off key."
From a heckler in the darkness there came a scornful "BOO!"
And a voice backstage caught Casey's ear and clearly said,
     "That's two."

Once more a grumbling started building in the room.
The patrons and the players thought the show had met its
     doom.
But with one stately gesture Casey stopped the growing
     groan,
When he pulled apart the curtains to reveal a giant throne.

Then from the wings there walked a damsel so petite,
And Casey, The Great Casey, gently placed her on the seat.
Now Casey threw a scarlet cloth deftly in the air
And soft as a fog it settled down and covered girl and chair.

Then Casey struck a pompous pose and pulled the silk aside,
And a puzzled frown creased every face...the girl still sat
     inside.
Casey's just not the same—he's lost everything they feared,
But Casey smirked, held up one hand...and then he disap-
     peared!

Oh, somewhere in this favored land they talk about a show;
There's delighted clapping somewhere, and you hear a loud
  "Bravo!"
And somewhere "Encore—Encore!" resounds from every
  nook;
But there is no joy in Vaudeville—The Great Casey got the
  hook.

Trivia

- 1 Year after *Casey at the Bat* author E. L. Thayer dies (1940),
  F. G. Thayer sells his magic business and home (1941) to William
  Larsen Sr. (Father of Bill Larsen and Milt Larsen of Magic Castle
  fame).

- 11 Years after Houdini, first president of SAM assembly #16 in
  Worcester, Mass. dies (1926), Ken Fletcher is born in Worcester,
  Massachusetts (1937). (Both miss SAM banquet that year.)

- 111 Years (to the day) after *Casey at the Bat* is first published
  (1888), *Casey at the Fox* is first published (1999).

- 1111 Days (3 years, 2 weeks, 2 days) after this announcement
  (March 4, 2003) James Randi will declare that the unusual numerical progression of these events is just another *coincidence.*

Sleight of Hand with Playing Cards prior to Scot's Discoverie

William Kalush

In 1584, Reginald Scot's seminal *Discoverie of Witchcraft*[1] offered for the first time in any language a full and detailed description of sleight of hand feats with playing cards. A surprising number of earlier mentions exist. From complete accounts with names, effects, and methods to merely a line here or there, they all allow the historian to better understand the state of the art prior to Reginald Scot. Techniques of both the card cheat as well as the conjuror will be considered.

The recorded history of sleight of hand magic begins at least as early as 2500 B.C.[2] Unfortunately, for our purposes here, playing cards don't make their appearance in the West until the third quarter of the four-teenth century. The first Western reference to playing cards seems to be from Spain in 1371.[3] It's likely that some clever but anonymous person soon-thereafter decided to use them in new and deceitful ways. Perhaps these deceits were an honest attempt to entertain and amuse, but equally as likely they were used to lighten another's purse. Less than forty years later, we find the earliest reference to sleight of hand with cards yet discovered.

William Kalush is a passionate researcher into the earliest conjuring history. He is currently working on a bibliography of European books pertaining to conjuring prior to 1701. The source materials for this article come from rare and valuable documents. For the enthusiast, direct quotes from these sources are included at the end of the article beginning on page 137.

[1]Scot, Reginald. *The Discoverie of Witchcraft* ... Imprinted at London by William Brome. 1584.

[2]Dedi of desdinefru performing for King Cheops. Westcar papyrus, Berlin State Museum.

[3]Parlett, David. *A History of Card Games,* Oxford, Oxford University Press, 1991. Page 35.

The earliest discovered record of deception with playing cards dates to 1408 France,[4] from a letter of remission preserved at the French National Archives.[5] The letter notes that several cunning men were caught using a less than honest stratagem (Quotation 1, page 137).

The ruse was basically a method by which Colin Charles and his partners would get other men to bet on what seemed to be a game of pure chance. The idea was that as cards were dealt around, the victim was to pick a certain card. Apparently unknown to M. Charles and his associates, the card the player was to find was subtly marked or smudged on its normally white back. Of course, the "sucker" would keep this "secret" to himself and surreptitiously win hand after hand. Certainly thinking that he had stumbled upon some of the most naïve men in town who couldn't even win their own game, he would gladly play and increase his wagers. Eventually he would bet all he could manage; the last hand would be dealt; and with confidence he would turn over the card with the subtle little secret mark. To his great chagrin, it would not be the card he was searching for. The pack had two such cards marked identically. Although the method our pioneer card cheats used to switch the cards is unknown, this certainly was the true point of the game. Having been caught at this con the men were tortured to confession and sentenced to the pillory. The letter of remission served to reduce their sentences.

This is quite a clever scam, made all the more impressive when its antiquity is considered. Notwithstanding, I am reminded of a point made to me by fraudulent gaming expert and consultant Steve Forte, regarding modern day card cheats. Upon my request to view some Las Vegas surveillance footage of discovered sharpers, Mr. Forte bristled and explained that only the poorest practitioners get caught; the best are never discovered. One must leave to the imagination what the true state of the art was in 1408. No memoirs of any card cheats from that time have yet been discovered.

In his very successful, off-Broadway, one-man show, *Ricky Jay and His 52 Assistants,* Ricky Jay has eloquently shown that some of Franois Villon's poetry[6] is a great source of information regarding late fifteenth century French lowlife. Since he was a thief and a lowlife himself, anything Villon has to say on the subject is greatly important. Villon's "La

---

[4] Allemagne, Henry Rene d', *Les cartes a jouer du XIV au XX siecle*, Paris, Librairie Hachette et cie, 1906. Also, Thierry Depaulis, "The Playing Card" Vol. X, no. 4. This was quoted from Hjalmar, "Le Bonneteau", published on the internet, http://rafale.worldnet.net/~fderik/LeBonneteau/LeBonneteau.html.

[5] JJ 162, ndeg361 (fdeg 264 Rdeg and Vdeg) citation from Hjalmar.

[6] Specifically "Tout aux tavernes et aux filles."

ballade des tireurs de cartes"[78] is a thirty- five-line song about cheating
at cards. Revealed are the concepts of hiding cards in a secret pocket
(*jabot*), conveying cards from this secret pocket to the hand, and the idea
of using marked cards (*pictonnez*). Similarly, though less significantly,
Eloy D'Amerval's *Le livre de deablerie*,[9] Paris, 1508, mention is made of
deceits and frauds with cards, but the details are not given.

In Milan before the close of the fifteenth century, Luca Pacioli, a
Franciscan brother in the Catholic Church, with the help of a young
artist, wrote what I consider to be the first book devoted primarily
to conjuring. The young artist was Leonardo da Vinci who, it can be
conclusively shown, had an independent interest in conjuring methods,[10]
but I must leave that topic for another time. Together the two wrote
De viribus quantitatis.[11] This marvelous manuscript, of which only one
contemporary copy is known, contains a great number of descriptions of
conjuring effects and concepts closely allied. At the end of Item XXX,[12]
which is a description of one of the many ways to divine a number or
numbers that are merely thought of by a spectator, is what may be the
earliest described method for a card effect. Pacioli is clear that this
method of number divination is intended to be performed for a group
of people, and that this group might request a repeat of the effect.
Knowing that repeating the same method often also exposes it, Fra
Luca explains an alternate method that would allow repetition without
fear of detection. He describes how the performer, prior to the event,

---

[7] Circa 1480.

[8] Fanch Guillemin, "Francois Villon et la Bible des Tricheurs", "Imagik", Numero
Special Hors Serie No. 3.

[9] *Le livre de la deablerie. L'imprimeur est Michel Le Noir, qui Paris a son
manoir. L'an mil cinq cens et huyt sans faulte.* [1508]. Noted by Mr. Guillemin,
see footnote 8.

[10] It has been Mr. Vanni Bossi's hunch for some time that Leonardo da Vinci had
some interest in conjuring. It should be mentioned that contained in the portion of
his notebooks that have been translated into English are several conjuring stunts. I
have also discovered that in the last century Gilberto Govi considered much of the
conjuring in De viribus to have been of Leonardo's invention. Also see footnote 18
below.

[11] MS 250, University of Bologna. First brought to my attention by David Singmaster. First to discover the conjuring connection was Vanni Bossi.

[12] *De viribus quantitatis* has only been printed once. *De viribus quantitatis / Luca
Pacioli; trascrizione di Maria Garlaschi Peirani dal codice n. 250 della Biblioteca
di Bologna; prefazione e direzione di Augusto Marinoni.* Milano : Ente raccolta
vinciana, 1997. I am indebted to Vanni Bossi of Legnano, Italy for bringing this to
my attention and in assisting me in obtaining a copy. Translated from the Italian by
Jeremy Parzen.

could teach a young boy a system that remains a secret between the
two. This system allows the performer to communicate information to
the boy using a variety of subtle methods. The performer might turn his
back and communicate through finger signals while his hands are behind
him, or, more subtly, he might resort to certain code words cloaked in
the guise of threats, such as "go back" or "go there." These words would
intimate certain information to the boy without allowing the audience
to know anything covert had been relayed.

Then our groundbreaking author touches on the idea of applying this
code to other objects, including playing cards:

> And you will teach the said lad, since he is closed up or far
> away, to guess which card they have touched with out seeing
> them when you have come with him for the numbers [trick];
> you will do so by assigning numbers to the figures and cards
> according to the tricks and according to the understanding
> between you; and you will give great pleasure to the group,
> for it will seem to whomever does not know the way, that
> all of this things have been done by [the] magical art [of]
> divination, etc. And thus for the points, the dice, and the
> ring, and the 3 varied things, and with him you will do stu-
> pendous things. But as I say, it is important that you do it
> very well, carefully, so that you will not be shamed, because
> these things are considered as secret as they are considered
> good.

I consider the description above to be the earliest card magic explanation yet discovered. As if this were not delightful enough, Pacioli
also relates that Giovanni de Jasonne of Ferrara used a similar method,
which Pacioli himself witnessed in Venice (Quotation 2, page 137).

Since the end of the eighteenth century, a similar method has been
used by magicians to perform what has become known as "second sight."
Interestingly, it can now be said that this method was used at least as
early as the end of the fifteenth century. What can not be said is what
might have happened to the literature of conjuring had this wonderful
manuscript been printed, allowing generations of conjurors access to
material that would lay dormant for centuries waiting to be rediscovered.
Perhaps if it had been printed, it would have dramatically increased the
quantity and the quality of what was subsequently written in regard to
sleight of hand in the sixteenth century.

The *Liber vagatorum*[13] [Augsburg, Joh. Froschauer, ca. 1509], a German booklet with a Latin title, is more or less a dictionary of slang used by vagabonds, criminals, and cheats. The reference to cards is only in passing, but in consideration of its age, it should be mentioned here.

> "*Item,* beware of the *Joners* who cheat at cards, who
> deal falsely and cut one for the other, cheat with *Boglein*
> and *Spies,* pick one card from the ground, and another from
> a cupboard..."[14]

Later our anonymous author says he might say more but "for your own good, I had better not explain." This line of reasoning is repeated continuously throughout the literature of cheating, up to the present century. Perhaps it was a real attempt to avoid luring the honest reader into the depths of depravity, but more likely, I suspect the author knew no more.

The above quote is the earliest mention of "dealing falsely" that I have been able to ascertain, but shortly thereafter buried in the famous and popular Il *cortegiano*[15] by Baldassarre Castiglione, there appears another. This entertaining guide to court life, written between 1508 and 1516, contains what may be the first appearance of an oft repeated practical joke. The gist of the conceit is that three men were playing cards (primero perhaps) and quickly one of the men lost all his money. Being a bit bored, the loser went to bed. The other two players decided to have a big laugh at their friend's expense and conspired to play a joke. They waited until they were sure that he was asleep. Then they quietly put out all the lights and curtained the window so no light at all could be seen. They took their places at the table at which they had been gambling and began arguing with each other so loudly as to deliberately wake their victim. Once awoken, their victim would not be able to see anything at all, but his mischievous friends would act as though they could see normally. They of course would convince the third man that the lights were lit and, if he could not see that, then the problem lay with him. He, believing himself to have gone blind, would panic, thus creating

---

[13] Cited by Kurt Volkmann in "The Origin of the Shift," *The Sphinx*, Vol. 51, No. 1, and brought to my attention by "Prose and Cons", a lecture by Ricky Jay, in the Pforzheimer Lectures on Printing and the Book Arts, at the New York Public Library, March 29, 1994.

[14] *The book of vagabonds and beggars; with a vocabvlary of their langvage and a preface by Martin Lvther; first translated into English by J. C. Hotten and now edited anew by O. B. Thomas.* London, The Pengvin press [1932].

[15] *Il libro del cortegiano.* Nelle case d'Aldo & d'Andrea d'Asola: Venetia, 1528.

the amusement his friends were looking for. One may wonder how has this anything to do with cards. The two things that the two pranksters said to each other to create the fight were "You've drawn the under card" and "And you have wagered on four of a suit."[16] Admittedly, this is a vague reference, and perhaps Castiglione had no direct knowledge of such artifice. He simply could have assumed that to cheat, one must deal from the bottom. Nonetheless it is worth mentioning here.

Hidden away in the British Library[17] is a well-preserved pamphlet of 4 leaves entitled *Opera nuoua doue facilmente potrai imparare piu giuochi di mano et altri giuochi piaceuolissimi & gentili come si potra legge[n]do uedere et facilmente imparare.* [G. S. di Carlo da Pavia: Florence, 1520?]. Although the anonymous author was generous enough to explain a dozen or so interesting feats of conjuring, he did not see fit to include any card magic per se. What he did, however, was include a warning to card players that others might use artifice to take the advantage, and therefore the money. Luckily enough for us, he described specifically several maneuvers with cards in an entry ostensibly to protect the player from fraud. Here he described how a cheat might use soap on desired cards to cause a subtle separation and thus find those desired. Also we are informed that a cheat might deal the second instead of the first card, and he might use a mirror to know what cards the others hold.

Although, as we have seen, the general concept of dealing falsely appears earlier in *Liber vagatorum* and the concept of bottom dealing crops up in Italy about the same time, the above passage is a clear exposé of what is still used and known today as a second deal. The reference to soap also may be the first time this ever-repeated stratagem appears. Soap allows the cheat to find desirable cards without having to resort to searching. Used until recent times, the soap (later wax) was applied to any desirable cards and would, upon slamming the deck down on a table or floor, cause the pack to separate at the soaped cards. The idea of using a mirror to see your opponent's hand is also mentioned repeatedly throughout the forthcoming literature, but this is the earliest mention I have yet found.

Although, strictly speaking it is outside the scope of this paper, it can be shown that mathematical "divination" stunts have existed in the West from at least the eighth century and can even be found hidden in

---

[16]Translated by Leonard Eckstein Opdycke. *The Book of the Courtier by Count Baldersar Castiglione,* London, Duckworth & Co., 1902.

[17]British Library C.20.a.31.(8.).

the personal manuscripts of Leonardo da Vinci.[18] What is now familiar to many as the "twenty-one card trick" does not seem to appear in print before 1593,[19] but the concept of mathematical card effects is mentioned in print seemingly for the first time in 1534. Franois Rabelais, in the first book of his important work *Gargantua and Pantagruel*[20] gives us our first glimpse. Unfortunately, he does not elaborate on what specific effects could be done. He says only:

> ...This done, they brought in cards, not to play, but to learn a thousand pretty tricks and new inventions, which were all grounded upon arithmetic. By this means he fell in love with that numerical science, and every day after dinner and supper he passed his time in it as pleasantly as he was wont to do at cards and dice; so that at last he understood so well both the theory and practical part thereof, that Tunstall[21] the Englishman, who had written very largely of that purpose, confessed that verily in comparison of him he had no skill at all. And not only in that, but in the other mathematical sciences, as geometry, astronomy, music, etc.[22]

Another brief reference to cards appears further down in the same section (Quotation 4, page 138).

———————

Although the literature of the sixteenth century is littered with mentions and detailed explanations of arithmetical divinations, the above passage is the earliest mention I have been able to find referring specifically to applications with cards. It is a pity we do not know which effects

———————

[18] See "Arithmetical Divination From Charlemagne's Court to Leonardo da Vinci." William Kalush, Atlanta 2000. Published for the Gathering for Gardner 4.

[19] Galasso, Horatio. *Giochi di carte bellisimi di regola e di memoria.* Venetia, 1593. This wonderful and extremely rare book has over 50 conjuring tricks, most of which seem to be new.

[20] *The most horrific life of the great Gargantua father of Pantagruel composed in days of old by M. Alcofribas abstractor of Quintessence book full of Pantegruelism.* On sale at Lyons, at Francois Juste's opposite our lady of comfort M.D. XLII.

[21] Tunstall Cuthbert 1474 1559.

[22] Translated by Sir Thomas Urquhart. *The works of Francis Rabelais, Doctor In Physick: Containing five books of the Lives, Heroick Deeds, and Sayings of Gargantua, and his Sonne Pantagruel. Together With the Pantagrueline Prognostication, the Oracle of the divine Bacbuc, and response of the bottle. Hereunto are annexed the Navigations unto the sounding Isle, and the Isle of the Apedefts: as likewise the Philosophical cream with a Limosm[!] Epistle. All done by Mr. Francis Rabelais, in the French Tongue, and now faithfully translated into English.* London, Printed for Richard Baddeley, within the middle Temple-gate. 1653.

Figure 1.   Arentino by Titian.

were done. Perhaps the elusive details will be discovered sometime in the future.

Famous Italian renaissance poet and author Pietro Aretino published his *Dialogo di Pietro Aretino nel qvale si parla del gioco con moralita piacevole*[23] for the first time in Venice in 1543. Considering that the entire book is a dialogue between a man from Padua and a playing card, I suspect careful study by those interested in the history of playing cards wouldn't be time wasted. I will refer to one passage only:

> PAD[OVANO]: As for ribaldry at cards.
>
> CARTE: Suffice it for you to understand that [there was] a Spaniard [who] used to carry inside his left arm a loose iron [rod]; as he would pick up [the card], it would come into his palm lengthwise; and as he would put his elbow down, it pushed out the card that came to him in the cut, pushing away the bad card in the hidden device with ability [worthy] of Spanish wool.[24]

This passage wonderfully describes what is now known as a "hold-out." This device, of which a great many designs and implementations are now extant, is a machine worn under one's clothing that switches a

---

[23]In Venetia: Per Giouanni de Farri, & fratelli, 1543. Brought to my attention by Gianni Pasqua of Torino, Italy.

[24]Translated by Jeremy Parzen.

card or cards secretly during a card game. Most commonly, as described by Aretino, it is in the cheat's sleeve. Through some innocuous action, it mechanically pushes the desired cards (previously placed secretly in the sleeve) into the palm. Then upon reversal of the procedure the dealt cards are taken away. Until this passage was discovered, the holdout was generally believed to be an invention of the late eighteenth or early nineteenth century.

———————

To the best of my knowledge specifics regarding cheating with cards do not appear in English until briefly touched upon in 1545 and then a bit more detail is lent to the subject about 1552, which is rehashed in 1577.

The first English reference considered here comes from a most unlikely source. Latin teacher to Queen Elizabeth among others, Roger Ascham published for the first time in 1545 *Toxophilus, the schole of shootinge*.[25] The work, ostensibly a manual of archery, is divided into two books; the first is what concerns us here. In a section derisive toward gambling (his reasoning is that it takes time from archery), Ascham finds place to briefly touch upon cheating with cards and dice. If what he says of dice is vague, then his mention of artifice with cards is minimal to the extreme. Unfortunately, he only sees fit to mention in passing the terms "false dealing, crafty conveyance, and false forswearing". Even though Ascham lumps cards with dice, he gives us just the smallest of tastes.

Next we find an exciting pamphlet, *A manifest detection of the moste vyle and detestable use of diceplay, and other practises lyke the same,...*[26] This anonymous pamphlet has been roughly dated to around 1552[27] and dubiously attributed to Gilbert Walker (G. W.), of whom nothing is known. Here we find somewhat more than we did in *Toxophilus*, and it is quite a bit more interesting (Quotation 5, page 138).

Here again we are informed of marked cards and cards deformed in such ways as to be found by the cheater using only the sense of touch. "Bum" cards can be narrow, wide, or perhaps bent. G. W. also notes

———————

[25] *Toxophilus, the schole of shootinge.* Londini: in aedibus Edouardi VVhytchurch, 1545.

[26] Imprinted at London, in Paules church yards at the sygne of the Lamb, by Abraham Vele.

[27] Many scholars find it likely that the only extant edition is either not the first, or earlier than the 1552 attribution. (There is rumor of both a 1532 edition and one with a printed date of 1552. It's questionable whether either has ever existed, and if so they both seem to be ghosts now.) From internal evidence it seems the book was written during the battle of Bolougne which ended in September 1544. Ricky Jay points out that upon a close reading it also sounds like *Toxophilus* is derivative of *Manifest detection* which would imply an earlier publication date to *Manifest detection* than formerly attributed.

how cheats might gain information by signals or mirrors. He tells of cards turned up (now known as a crimp or wave), and even more interestingly, he mentions "prick." Prick could be cards marked by using fine needles to make bumps on the cards that the cheater could feel. It is also possible that the author had another meaning for the word that is now obscure. Interestingly, G. W., like Aretino before him, mentions "one fine trick brought in [by] a Spaniard." This topic will briefly be covered later. John Northbrooke's *A treatise against dicing*[28] followed in 1577, but was completely derivative of what had come before.

In stark contrast to the sparse English offerings from the mid sixteenth century is: *Le mespris & contennement de tous ieux de sort compose Oliuier Gouyn de Poictiers.*[29] It is a beautiful book printed in Paris in 1550, that gives marvelous details regarding cheating with cards. Although I can say nothing of Olivier Gouyn, I can say that this book (I believe the only one he wrote) is a wonderful contribution to the history of chicanery we are now examining. His third chapter is devoted entirely to methods of cheating at one game or another. Entitled "Of the subtleties, ruses, deceits, cheating and nasty things that are done in games," he devotes about a third of the twenty-four, 12deg pages of the chapter to artifice with playing cards. He warns repeatedly that you, the ordinary player, cannot hope to win against a professional who cheats. If you attempt to protect yourself by requesting a new pack, the cheater can mark the cards while playing. Or, he will send someone he knows and is in confederacy with who will bring back a new looking pack that has been gimmicked in several ways to allow the advantage to go to your adversary. He also tells that if you buy the cards yourself, no matter, the cheater can switch the pack at any moment. Perhaps you will spit on the floor or blow your nose; when you look back, the cards will have been switched. The pack can be marked beforehand, or even the size of the cards can be altered. Long, wide, short, or narrow cards are all mentioned. This subterfuge would allow the cheat to gain control over desirable cards, thus causing them to be dealt or avoided, as need dictates. Gouyn also warns that these insidious players will also pair

---

[28] *Spiritus est vicarius Christi in terra. A treatise wherein dicing, dauncing, vaine playes or enterluds with other idle pastimes &c. commonly used on the Sabboth day, are reproved by the authoritie of the word of God and auntient writers. Made dialoguewise by John Northbrooke.* London, H. Bynneman for Goerge Byshop, [ent. 1577].

[29] Listed by Manfred Zollinger in his *Bibliographie der Spielbücher des 15. bis 18. Jahrhunderts,* Anton Hiersemann Verlag, Stuttgart 1996. This was brought to my attention by Daniel Rhod of Paris.

up and cheat using signals and even affix mirrors to their clothing to allow secret reading of an opponents hand. Cheaters also have ways to seemingly shuffle the cards but actually keep them in the same order, especially those underneath. "And after you have cut (if you don't pay too much attention), he will put underneath, that which should go on top." And when dealing, the cheater will deal from the bottom or the middle. We are told that the marked cards can be held back by the dealer so they do not go to his opponent and end up in the dealer's hand. Regarding the level of skill these men possess Gouyn says: "For never has a juggler[30] had more supple hands and fingers playing the cups,[31] than some cheaters have in handling cards and dice." The remainder of the chapter concerns cheating at dice and bowling, interesting topics that we must not delve into now. This wonderful book breaks new ground in exposing the existence of several deceptions that I believe heretofore were usually considered much later inventions.

To the best of my knowledge, Girolamo Cardano's *De subtilitate*[32] has the honor of containing the first printed description of a method for a card effect (Quotation 6, page 139). (Bearing in mind that Fra Luca's manuscript was never published.) In his effect he explains two methods to retain knowledge of where the card is. First by using the finger, a topic later expanded by Scot, and now known as a "break". The second is by placing the selected card in close proximity to a known card. This would allow the performer to shuffle to some extent and still be able to find the selection.

Cardano was a Milanese physician who, fortunately for us, had two pastimes. He gambled, and he wrote about it. Although there is more magic revealed in his *De rerum varietate*[33] than in any of his other works, his famous *De subtilitate* and his contemporaneously unpublished *De ludo alea* will concern us here.

Kurt Volkmann in "Magie" first noted that the beginning of Book 18 of *De subtilitate* contains a long description of the performance of two magicians, Dalmagus and Francesco Soma (Quotation 7, page 139). Since no card magic was attributed to Dalmagus, I will not deal with him here. Francesco Soma of Naples, however, is a different matter.

---

[30] Juggler and its variant spellings refer to what is know called a magician, and not necessarily one who can keep several balls in the air at one time.

[31] Referring to the classic cups and balls.

[32] *Hieronymi Cardani mediolanensis medici de subtilitate libri XXI.* Norimbergae: Petreium, 1550.

[33] *Hieronymi Cardani mediolanensis, medici, de rervm varietate libri XVII.* Basel, Henric Petri, 1557.

The first edition of *De subtilitate* was published in Nuremberg in 1550,[34] and lacks any mention of Soma whatsoever. The anecdote first appears in the edition of 1560, which is also the last edition published during Cardano's life. From this we can deduce that Cardano likely witnessed this performance within this 10 year window.

According to Cardano, Soma could reveal a selected card even though it hadn't been returned to the pack. Soma could also cause several participants to select the same card. To explain this, Cardano considers the possibility of a pack consisting solely of one value and suit of card. But this theory he realizes is flawed and he rejects it. When Cardano invites an Epicurean philosopher to scrutinize Soma, the man notices that the conjuror is "murmuring" throughout his performance. This suggests to me that perhaps Soma was able to successfully apply mnemonic techniques to his conjuring methods. This conjecture might not be unfounded considering the publication of such a concept as early as 1638.[35]

Unpublished until Cardano's complete works were printed in 1663,[36] *De ludo alea* also mentions Francesco Soma, but this time the spelling is Sorna. I suspect this as an error in transcription of Cardano's manuscripts, but I must admit that without checking I cannot say for certain. As the title implies, this book is devoted to gambling games. In Section 17, Cardano briefly explains some of the then current methods used to cheat at cards. In light of earlier material quoted above, it should suffice to say that Cardano briefly covers the concept of marking the cards in several ways and touches on the idea of mirrors and secret codes to gain knowledge of an opponent's cards. Notwithstanding this, there is one curious comment. Near the end of this section, Cardano writes:

> Since prestidigitators are capable of such admirable feats, why is it that they are usually unlucky at cards? It would seem reasonable that, just as they are able to deceive us with balls, pots, and coins, they should also be able to do it with cards and so invariably come out winners. But the condemned Spaniard was ordered (in fact, the prohibition, they say, was on pain of death) not to play, seeing that he could at will produce four cards that make chorus either by

---

[34] This edition mentions a Spanish magician in the retinue of Charles V, but Dalmagus's name is not present. It appears at least by the 1554 edition.

[35] Siviero da Cento, Benedetto (Il Carbonaro). *Nova Ghirlanda di bellissimi giochi di carte, e di mano. Con altri bellissimi gioch d'intertenimento.* Data in luce da me Benedetto Siuiero da Cento detto il carbonaro. Venetia, Fiorenza, Bologna, Oruiero, Padoa e Macerata, Perugia & in Roma, appresso Bernardino Tani, 1638.

[36] *Opera omnia: tam hactenvs excvsa;* Lvgdvni, Sumptibus Ioannis Antonii Hvgvetan, & Marci Antonii Ravavd, 1663.

quickness of hand; for we must assign to either of these a prodigious art of prestidigitation.[37]

The inferences in the above passage are quite interesting. Cardano implies that not only has he had much contact with prestidigitators, but it's to the extent that he has played cards with a great enough number to generalize on their relative luck. From personal experience I can say that even a great amount of skill at performing card effects relates very little to the same performer's ability either to play cards well or, for that matter, to cheat. Although it can be said without question that many of the same sleights used to cheat at cards are also used in performing conjuring feats, the similarity ends there. The moment when the crucial sleight can be performed is almost always completely different when used for cheating as opposed to entertaining. Also I might point out his reference to a condemned Spaniard. I must confess that I have been unable to find out who this Spaniard was, my only guess being that it could have been the same Dalmagus he mentions in De subtilitate. I suspect a Spanish scholar might be better able to locate more information regarding Dalmagus. I have been unable to find anything further than what Cardano himself said.

An interesting aside to this chronicle is the use of playing cards to relay secret messages. Anticipating by 400 years a card effect somewhat popular in the twentieth century, Gianbattista della Porta of Naples published an important work on secret writing in 1563.[38] He included a vast number of avant-garde techniques to communicate secretly, one example using nothing more than a pack of cards. The two parties wishing secure communication need only agree on a specific order of an ordinary pack of cards, also agreeing on which, if any, of the cards would be face up and which would be face down. Once the two parties each know the secret order, the sender places his pack in this order then writes his secret message on the edges of the pack. When the sender finishes writing the message, the sender then shuffles the deck thoroughly and randomly, turning some cards over and leaving some the way they were. The shuffled pack is then transported to the party needing the secret information. The concept is simple and quite brilliant; if any interceptor wants to read the message, the pack must be reassembled. The number of distinct orders that a single pack of playing cards can occupy is

----

[37] Translation from *Cardano, the gambling scholar.* Oystein Ore, Princeton, Princeton University Press, 1953.

[38] *De furtiuis literarum notis: vulgo, De ziferis libri IIII / Ioan. Baptista Porta Neapolitano autore.* Imprint Neapoli: Apud Ioa. Mariam Scotum, 1563.

absolutely staggering. With the aid of Professor David Singmaster of London, I determined that an ordinary pack of 52 cards can be put in 52! distinct orders which is equivalent to 8.1×1067. If the wrinkle of face up and face down is considered, this number rises to 3.6×1083, which I believe to be about equal to the number of atoms in the universe! I suspect this method of transfer of secret information may have failed to catch on for two reasons: (1) The amount of information is limited, due to lack of writing space and (2) the vulnerability of transferring the key. This form of cipher apparently resurfaced in the twentieth century, to be used in World War One. This lead to a challenge posed by Theodore Annemann in *The Jinx*, No. 19, April 1936. "Somebody, somewhere, may make use of this idea. During the war a code was intercepted which used a deck of cards. They were in a certain order and the message was written on the edge of the deck while it was gripped or tightly held. Then the cards were shuffled. Only the person knowing the order could put them together to make the message readable." In response to this challenge the concept was applied and has become the root of several entertaining mysteries now used by modern card conjurors.

Girolamo Scotto[39] was an Italian performing card magic in the fourth quarter of the sixteenth century. Surprisingly, a great deal is known about his most interesting life. Scotto was reputedly a Knight of Piacenza and is the first card magician of which a representation of his face is known. Scotto was attached to several European courts, not the least of which were Ferdinand II's of Vienna and Rudolpho II's of Prague. Credited to Scotto is the ability to take four cards from a pack and make them change repeatedly to other cards. He would ask the first spectator to name any four of a kind. The first spectator says kings, and immediately the four cards become kings. The next spectator requests queens, and the same four cards are shown to have become queens. The third says aces, and the cards are aces. Finally the last person, being a bit belligerent, says "nothing." Scotto relents and shows that the cards have now become blank.

Also credited to Scotto is the ability to cause a spectator to select the same card over and over again. Perhaps his method might have been similar to that used by Francesco Soma. A detailed first hand account has survived (Quotation 8, page 140).[40]

---

[39]First mentioned by Sidney Clarke in "The Annals of Conjuring"; see full citation in footnote 46. Full articles followed by Ottokar Fischer, "Hieronymus Scotto, An Unknown Conjurer of the Renaissance", *The Sphinx* Vol. 36. Edgar Heyl, "New Light on the Renaissance Master", *The Sphinx*, Vol. 47.

[40]Reported contemporaneously by Archduke Ferdinand's physician, Dr. Handsch. Hirn, Josef, *Erzherzog Ferdinand II. von Tirol; Geschichte seiner Regierung und seiner Länder.* Innsbruck, Wagner, 1885–1888.

Figure 2.   Medal of Scotto by Abondio.

Newly discovered by collector and historian Giovanni Pasqua of Torino, Italy is a pamphlet entitled *Secreti di natura maravigliosi del sig. Gieronimo Scotto Piasentino.*[41] This exciting find consists of a modest eight pages and contains ten items, five of which are card effects. Without date or printer, it is difficult to know when this booklet first appeared, but I suspect it was contemporary with Scotto.

The tract explains,

> [How to] make cards walk, dance, and how to pull out the one you want from all the others.
> [How to] make primero from four cards and [how to] make a flush from primero and [how to] make four [no trumps?] out of a flush.
> [How to] make a card of spades turn into a [card] of diamonds or [how to make] goblets into coins.
> [How to] guess cards while blindfolded.
> [How to] guess which card imagined by one or two [and] if they are truly taken, you will [even be able to] guess the card [held by all] three.

---

[41] The only known copy is in Mr. Pasqua's private library.

Considering that at least one of these items is precisely what has been independently attributed to Scotto, and one of the others is very similar, it is not reckless to speculate that Scotto may have done the other effects described. Whether or not the methods are the same as Scotto's is a question we might never answer for certain. I can say, however, that all of the methods for card effects described in the libretto have come down to us and have been in use, in one form or another, up to and including the present day.

———

Scotto's contemporary, amateur conjuror Abramo Colorni,[42] was a clever engineer who is remembered mostly because he was both Jewish and in high standing at several European courts, including that of the Duke of Ferrara. At that point in European history, being of the Jewish faith could be quite detrimental to one's societal status. Colorni stands out as an exception to that rule. We are interested here in his conjuring abilities with cards, of which, fortunately, a little bit is known. Although he was mentioned by several contemporary authors, and some of his personal letters survive, the most important accounts regarding his magic were written by his friend Tomaso Garzoni. Garzoni's most famous and important book *La piazza universale di tutte le professioni*

———

[42]Kurt Volkmann, *Magie* 1942, and Robert Lund, "Colorini", *The Sphinx* Vol. 52, No. 1.

del mondo[43] contains a nice account of Colorni's card conjuring (Quotation 9, page 141). Colorni is credited with being able to have a person select two cards, while still in the possession of the spectator one card changes into the other and then the surface of this card is scraped off to reveal some tiny writing which turns out to be the very thoughts of the person holding it!

We are also told Colorni could place a pack in the center of a table. Those around it could take whichever card they wished and the card would be the one he said. Colorni was also mentioned again by Garzoni in his much rarer and arcane *Il serraglio de gli stupori del mondo*, but the passage is nearly identical to that of *La piazza*. It has been interpreted by others, and I'm inclined to agree, that the passage "he makes any card named ... come out of the pack" is a direct reference to what is now popularly known as "the rising cards," a classic of card magic. The rising cards is most impressive when any card freely named by an audience member spontaneously rises. Alternatively, the first card item in *Secreti di natura maravigliosi* cited above is a method to get a card to dance or come out of the pack. This is not the rising cards, and is closer in spirit and method to what is now known as "the haunted pack." It is possible that the eyewitness accounts of Colorni were less than perfect in reporting the minute details of the effect. Magicians have long relied upon the audience mis-remembering what actually happens during a performance. For this reason, accounts of effects by nonmagicians can often be suspect.

The fact that Colorni was an engineer of great notoriety implies that he may very well have invented his own methods for his effects. Unfortunately, none of his secrets have yet been discovered. Like Gianbattista della Porta, Colorni's interests also included cryptography. He published *Scotographia*,[44] in Prague, 1593, and it was dedicated to Rudolph II, the same who had hosted Scotto at one time.

---

[43] *La piazza vniversale di tvtte le professioni del mondo, e nobili et ignobili / Nvovamente formata, e posta in luce da Tomaso Garzoni da Bagnacauallo.* In Venetia: Appresso Gio. Battista Somascho, 1585.

[44] *Scotographia overo, Scienza di scrivere oscvro, facilissima, et sicvrissima, per qual si uoglia lingua; le cui diuerse inuentioni diuisi in tre libri, seruiranno in piu modi, & per cifra, & per contracifra. Le qvali, se ben saranno commvni a tvtti, potranno nondimeno usarfi da ogn' uno senza pericolo d'essere inteso da altri, che dal proprio corrispondente. Opera di Abram Colorni ....* In Praga presso Giouani Sciuman. M.D.XCIII.

Lastly, Francis Bacon, who was known as the most learned man in Britain, included in his *Sylva Sylvarum*[45] a detailed account of a conjuror performing a card effect that doubtless would still be quite impressive today (Quotation 10, page 141).

Bacon relates that while living in his father's home (this would be around 1575[46] he had the opportunity to see a juggler, who performed a card effect. Later Bacon tells the story of this performance to a vain man he calls a "Pretended Learned Man." The effect was that one spectator was asked to select a card and another onlooker was to name any card that came to mind. The selected card and the thought card were then discovered to be one and the same. Bacon's interlocutor was of the opinion that the juggler had forced his will upon the man who had named his thought card. On this point I agree. In fact the juggler must have forced his will on both participants, but by different means. This concept of forcing is implied in the work of Francesco Soma as well as Girolamo Scotto. Interestingly Reginald Scot explained how both versions of this technique can be accomplished, leaving the juggler's effect exposed. Due to the countless redescriptions over the last 400 years, this technique has never gone out of fashion. It consequently has become one of the staples of all card conjuring to this day.

Without doubt the invention of sleight of hand maneuvers with cards was concurrent with the invention of the cards themselves. I suspect that there are countless mentions and descriptions of methods and performances that have yet to be uncovered. It is not reasonable to think that at anytime since the introduction of cards that conjurors' or sharpers' minds or hands have been idle. It is likely that the vast majority of the repertoire of modern card conjurors was already being used prior to the publication of Scot's *Discoverie of witchcraft*.

I sincerely thank all of the researchers who have come before me, upon whose shoulders I now stand. I also thank Marsha Casdorph for her fine help in editing. Also a large thanks to: Vanni Bossi, Ricky Jay, David Singmaster, Gianni Pasqua, Daniel Rhod, Ariel Frailich, Jeremy Parzen, and Marianne Santo.

---

[45] *Sylva sylvarum, or, a naturall history in ten centuries / by the Right Honble. Francis Lo Verulam Viscount St. Alban;* published after the authors death by William Rawley London: Printed by J.H. for W. Lee ..., 1626.

[46] Sidney W. Clarke, "The Annals of Conjuring" published serially in *The Magic Wand*, v. 13–17 (nos. 121–140), 1924–1928.

Quotation 1 (page 120, footnote 4):

[. . .] Charles VI, King of France by the grace of God, makes it known, to all who are in the present and in the future: we have received the humble plea of Colin Charles, of Montdobleau, in the bishopric of Chartres, poor young page, unmarried, helper to a mason, aged approximately 24 years, saying that he, in the company of 6 other companions, went playing above the Pont Neuf in Paris, about one year [ago], encountered two merchants from the country of Brittany. To one of which, one of the companions made it known, that he had frans [Francs] on his and a big silver tournament that he wanted to sell and asked him if wanted to buy them, saying that he would sell them and they went to the Bonne tavern, and he showed them and would give Francs for 16 solz parisis [currency]. And if said merchant were to pay, he would not owe any more; and so much did he and his companions ≪ennorterent≫ said merchant that he went with them and when they were in the tavern and had drunk, one of said companions reached for a number of papers to play [playing cards] and made said supplicant and his companions play said merchant who, by their seduction, played at guessing which card one would touch. So much so that, when he played for nothing, he won because he was shown[47] how the card he had to pick was marked. But they had not showed him that there were two cards marked in identical fashion, and suddenly, ≪par enviz er renviz≫ at every card picked as he was told, and of each consent, said merchant lost 22 écus [coinage] for not having picked the one he was shown, with a painting of roses on the front, but a similar one with the back marked as said. For this deed said Colin and others of his company were soon after caught and imprisoned in our Chastelet de Paris [small castle of Paris]. And for what was found to be matters of abuse and deception, said supplicant was very questioned in a hard manner, and, finally, his confession heard, was sentenced to the pillory and was put there, and banned from our Kingdom forever, and would never dare converse there without our grace [. . .].[48]

Quotation 2 (page 122, footnote 12):

With similar ways a certain Giovanni de Jasonne from Ferrara would perform. He had a lad that he had instructed since the cradle in similar gentilities: by means of numbers and by acts, signs and gestures of the hands, feet, and by coughing and by yelling, by beating knives on the table etc., he could make him understand 2 or 3 words that he would secretly say; he would have him guess by composing letters with space between words and stories, so that the group would not understand. And at the same time, because it seemed that the deed was not [just] his, he had ordered the lad to always keep his eyes on

---

[47] Although the manuscript uses the French word for "shown," I believe that for the ruse to work it must be assumed that the detection of the markings was incumbent upon the sucker.

[48] Translation by Ariel Frailich.

his hand, and thus he could compose words and syllables and numbers, etc., and with this he went often to Venice—and I was there—and he would do similar effects in the home of some gentleman or gentlewoman in such a way that they swore that the lad had a friendly spirit that revealed all of these things, etc.; and with these he went away... Finally the boy died but he is still here.[49] Similarly, you can use this ways and turns to teach one for the pleasure of refined men, as he did, and when he did such things, he was careful that there was no one that seemed to understand [...]

Quotation 3 (page 124, footnote 17):

If you want to play at «chiamare» every time you will give the second card and you will know the point that your partner is asking and that point [card] put a little bit of soap on that card. Square the cards on the floor and the card will come out and you will take another card and you will give them second and to play at «susso» or «premera» & you will have a mirror between your legs and you will know the card you are dealing.[50]

Quotation 4 (page 125, footnote 22):

After that they had given thanks, he set himself to sing vocally, and play upon harmonious instruments, or otherwise passed his time at some pretty sports, made with cards or dice, or in practising the feats of legerdemain with cups and balls. There they stayed some nights in frolicking thus, and making themselves merry till it was time to go to bed;

Quotation 5 (page 127, footnote 26):

R. asks: Then am I sufficiently lessoned for the purpose. But because at the first our talk matched Dice and cards together, like a couple of friends that draw both in a yoke, I pray you, is there as much craft at cards as ye have rehearsed at the dice?

To which M. replies: Altogether, I would not give a point to choose; they have such a sleight in sorting and shuffling of the Cards that play at what game ye will, all is lost aforehand. If two be confederate to beguile the third, the thing is compassed with the more ease than if one be but alone. Yet are there many ways to deceive. Primero, now, as it hath most use in court, so is there most deceit in it. Some play upon the prick, some pinch the cards privily with their nails, some turn up the corners, some mark them with fine spots of ink. One fine trick brought in a Spaniard; a finer than this invented an Italian, and won much money with it by our doctors, and yet at the last they were both overreached by new sleights devised here at home. At trump, sant,

---

[49]Pacioli's meaning here is not clear.
[50]Translated by Gianni Pasqua, Vanni Bossi, and Jeremy Parzen.

and such other like, cutting at the neck is a great vantage. So is cutting by a bum card finely, under and over, stealing the stock of the discarded cards, if their broad laws be forced aforehand. At decoy they draw easily twenty hands together, and play all upon assurance when to win or lose. Other helps I have heard of besides, as to set the cozen upon the bench with a great looking-glass behind him on the wall, wherein the cheater might always see what cards were in his hand. Sometimes they work by signs made by some of the lookers-on. Wherefore, methinks this among the rest proceeded of a fine invention: a gamester after he had been oftentimes bitten among the cheaters, and after much loss, grew very suspicious in his play that he could not suffer any of the sitters-by to be privy to his game. For this, the cheaters devised a new shift. A woman should sit sewing beside him, and by the shift or slow drawing her needle give a token to the cheater what was the cozen's game. So that a few examples instead of infinite that might be rehearsed, this one universal conclusion may be gathered: that give you to play and yield yourself to loss.[51]

Quotation 6 (page 129, footnote 32):

The manner of knowing the card that one has marked is as follows. Have someone visualize it in his thoughts, then show the cards one at a time: when he will indicate that it's the one he [lit: hears; thinks of?], secretly you will mark it with your finger and «incontinet» you will shuffle the cards, when you come across it, you will show it. Others put it in front of another (card) that they know [ie, key cards], and shuffle it with the others, and see it and know it before it is separated from the others, then they remove it or have it removed as their companion desires. Others find it by number, by often dividing the cards [piles?].[52]

Quotation 7 (page 130, footnote 34):

"Recently I made the acquaintance of Francesco Soma of Naples, a young man of high birth who by common account had scarcely reached his twenty-second year, and who, in addition to such exceptional skill in music that he has no equal in playing the lute, knows many incredible tricks of legerdemain. Among others there is one which I often witnessed together with my friends, and for which I have never been able to find a natural explanation.

He spread the cards on the table in such a way that the pack was not separated, and then he asked us to take one card and conceal it. Then he

---

[51] Text taken from *Rogues, Vagabonds & Sturdy Beggars*, edited by Arthur F. Kinney, Barre, Massachusetts, Imprint Society, 1973.

[52] Translated from the French by Ariel Frailich. This passage has been translated from the French edition of 1578, which had been translated from Latin by Richard Le Blanc, Paris 1556. I must admit that it would have been preferable to have worked with the original Latin, notwithstanding that even if Le Blanc changed something it is still middle 16th century.

took the pack, shuffled it, and guessed what card it was. This might perhaps have been attributed to quickness of hand; but that is by no means the case with what he did next. For when the card was put back in the pack and it was laid on the table, he asked several of us to draw a card, and we realized that the man who had drawn the card on the earlier occasion always drew the same card now, as though Soma were compelling us to draw the same card, or else were changing the face of the card.

When I brought a clever man, an Epicurean philosopher, to the spectacle, he confessed that he could not discover how it was done, although he did not for that reason think we ought to admit that there is any power in these sublunary things beyond that which we see granted to them by nature. Thus this man evaded all our watchful care and surpassed us in cleverness. While he was doing the trick he kept murmuring something constantly, as though he were calculating; yet it was certain that what he said did not consist of any reckoning with numbers. But when that well-known friend of ours, after taking a card, looked at it before putting it under a book, Soma said, 'You have confused everything and have spoiled my whole method; nevertheless, the card is the same as the one you drew before, namely, the Two of Flowers [clubs],' and we discovered that this was so.

And although he showed me certain more wonderful things, still indications were that all of them were the work of a certain art of legerdemain rather than of supernatural beings, or in other words, they were much less miraculous. Nevertheless, the art was too wonderful to be understood by human cogitation. And if he had not asked us at various times to draw different cards, I would have suspected that he had substituted a pack consisting of cards of a single kind, namely, the 'Two of Flowers.' For with that device it would happen that whoever drew a card would always seem to chance upon the same card. But, as I have said, the diversity of the remaining cards precluded that explanation.[53]

Quotation 8 (page 132, footnote 40):

I thought of a card, the eight of hearts, and he showed it to me. Each of the ten people present was made to remember a card; Scotto had cards drawn and, O miracle! Each one drew the card he was thinking of. Also he put four cards, which one had seen, into one's hand. Then he asked, 'What do you want?' and whatever you wanted, that was it. [...]When he repeated this experiment with the four cards, one of the gentlemen tried to embarrass him by answering to the question whether it should be kings, queens, or jacks, 'Nothing'. But at once all the cards were blank.

---

[53]Translation from *Cardano, the Gambling Scholar*. Oystein Ore, Princeton, Princeton University Press, 1953.

Quotation 9 (page 135, footnote 43).

Sometimes he will do the following experiment, viz., he makes someone take two cards, and tells him to imagine that one of the two is changed into the other, which punctually takes place! Again, stripping off the surface of a card, he will show written beneath in minuscular writing the very thoughts of the person who was holding it, or had it concealed in his bosom. At other times, he makes any card named by one of the company at will come out of the pack. He knows also a thousand other tricks of the sort, which by his kindness I have seen with my own eyes, together with more than ten other friends, who have all been equally amazed.

Quotation 10 (page 136, footnote 45).

For Example; I related one time to a *Man*, that was Curious, and Vaine enough in these Things; *That I saw a Kinde of* Iugler, *that had a Paire of* Cards, *and would tell a* Man *what* Card *he thought*. This Pretended Learned Man told me; It was a Miſtaking in Me; For (*ſaid hee*) it *was not the* Knowledge *of the* Mans Thought, *(for that is Proper to* God,) *but it was the* Inforcing *of a* Thought *vpon him, and* Binding *his* Imagination *by a* Stronger, *that he could* Thinke no other *Card*. And thereupon he asked me a *Queſtion*, or two, which I thought he did but cunningly, knowing before what vſed to be the *Feats* of the *Iugler*. Sir, (*ſaid he*,) *doe you remember whether he told the* Card, *the* Man *thought*, Himſelfe, *or bade* Another *to tell it*. I anſwered (as was true;) *That he bade Another tell it*. Whereunto he ſaid; *So I thought*: For (ſaid he) *Himſelfe could not haue put on ſo ſtrong an* Imagination; *But by telling the other the* Card, *(who beleeued that the* Iugler *was ſome Strange* Man, *and could doe Strange Things,) that other* Man *caught a ſtrong* Imagination. I harkened vnto him, thinking for a Vanity he ſpoke prettily. Then he asked me another *Queſtion*: Saith he; *Doe you remember, whether he bade the* Man *thinke the Card firſt, and afterwards told the other* Man *in his Eare, what hee ſhould thinke, Or elſe that he did whiſper firſt in the* Mans Eare, *that ſhould tell the* Card, *telling that ſuch a* Man *ſhould thinke ſuch a* Card, *and after bade the* Man *thinke a* Card? I told him, as was true; *That he did firſt whiſper the* Man *in the Eare, that ſuch a* Man *ſhould thinke ſuch a* Card: Vpon this the *Learned Man* did much Exult, and Pleaſe himſelfe, ſaying; *Loe, you may ſee that my* Opinion *is right: For if the* Man *had thought firſt, his* Thought *had beene* Fixed; *But the other* Imagining *firſt, bound his* Thought. Which though it did ſomewhat ſinke with mee, yet I made it Lighter than I thought, and ſaid; *I thought it was* Confederacie, *betweene the* Iugler, *and the two* Seruants: Though (Indeed) I had no Reaſon ſo to thinke : For they were both my *Fathers* Seruants; And he had neuer plaid in the Houſe before. The *Iugler* alſo did cauſe a *Garter* to be held vp; And tooke vpon him, to know, that ſuch a *One*, ſhould point in ſuch a *Place*, of the *Garter*; As it ſhould be neare ſo many *Inches* to the *Longer End*, and ſo many to the *Shorter*; And ſtill he did it, by *Firſt Telling* the *Imaginer*, and after *Bidding* the *Actor Thinke*.

Cubist Magic

Jeremiah Farrell

Color Plate IX shows a mystic 4×4 colored array of letters labeled the "Multidimensional Gardner" (or MG for short). The MG serves as a backdrop for an interesting, easy to execute, mathematical magic trick whose effect is as follows. Suppose we ask our friend Mark to secretly choose a single letter from the word ASTEROID. Suppose also we allow him the privilege of deciding (privately) either to be "convivial" and always tell the truth, or to be "contrary" and always lie. This last choice becomes his "quirk." Then we ask him four questions, to be answered according to his secret quirkiness.

RED question: "Is your letter in the word SEAT?"

BLUE question: "Is your letter in the word SOAR?"

YELLOW question: "Is your letter in the word RITA?"

GREEN question: "Is your letter in the word OTIS?"

After Mark's replies we immediately name his letter—and his quirkiness—with a mere glance at the MG!

This trick, once learned, can be repeated as often as one likes. And if the four questions are written on appropriately colored slips of paper, a variation of the trick can be performed that may seem even more dumbfounding. Let Mark silently separate the slips into two piles, one pile for those slips for which his response would be yes and one for which his response would be no. Even though we do not know which pile is which, we can still use the MG to quickly identify his letter!

The mathemagician **Jeremiah Farrell**, ex-newspaperman, ex-engineer, ex-planetarium director, and ex-professor, finally finds his true calling by heeding his mentor Martin Gardner.

Of course the colors are crucial in the performance of the trick, but before we can explain the inner workings of the mysterious MG, we must explore the underlying geometry of the situation.

Color Plate X contains diagrams of five multidimensional "cubes." The fifth "cube," and the one we are most interested in, is the four-dimensional tesseract consisting of 16 nodes labeled with red or blue letters from the word ASTEROID.

Notice that all horizontal edges are RED; all vertical edges are BLUE; all third dimensional edges are YELLOW; and all fourth dimensional edges are GREEN. These four edge colors correspond exactly to the colors of our four questions in the prediction tricks.

In fact, the tricks can be performed equally well on the tesseract as on the MG. Here is how the original trick is accomplished on the 4-cube. First we note which colors get yes answers from Mark. Then, starting from the blue D node in the lower right, we traverse those same colored edges (in any order) and arrive at a new node. The letter on the new node will always be Mark's chosen letter. If it is red, Mark is contrary; and, if it is blue, he is convivial.

For example, suppose Mark said yes to the RED, YELLOW, and GREEN questions. From the blue D we mentally travel, say, YELLOW to the red S, GREEN to the blue I and finally RED to the blue T. So Mark convivially had chosen T. Any other order of the three colors would also have ended on the blue T. The reader may check that choosing T and telling the truth does indeed elicit yeses to the three colors mentioned. Notice also that if Mark's letter was T and he decided to lie, then the only yes response would be to the BLUE question and we would thus travel the BLUE edge from the blue D node and end on the red T as required.

Since changing quirkiness on a given letter always gives complementary responses, we can effectively perform the second version of the trick where Mark separates his yeses and noes into two piles without telling us which is which. We simply follow either set of colors and will end on Mark's letter—but we cannot be sure of his quirkiness.

It is of some geometrical interest to note that we can discard the GREEN question and revert to three dimensions, where the trick will still work with the only proviso that Mark must now always tell the truth. We ask him the three remaining color-questions and traverse those colored edges on the ordinary cube (from the D as before) for which he responded yes and end on his letter.

For two dimensions we ask Mark to choose one of the four words, AS, ET, OR, or ID. He then answers, truthfully, the questions:

RED question: "Are your letters in the word SEAT?"

BLUE question: "Are your letters in the word SOAR?"

Start now on the lower right ID and his yeses direct us to his chosen word.

One dimension is trivial. Mark selects one of the words SEAT or DIOR and we ask him the RED question and immediately know his choice.

A zero-dimensional "question" is one that is never asked. We offer the letters RADIO SET to Mark and simply inform him that his letters transpose into ASTEROID.

The geometric content of each of these five multidimensional cubes is completely contained in the arcane MG diagram. In fact the MG can be imagined as a torus, or doughnut shape, by bending around the top and bottom half-red edges to form a tube and then joining the half-blue ends to complete the torus. Thus each one of the 16 nodes (i.e., letters) is connected to four others via exactly one of the four colors. For example, starting at the blue D we arrive at the blue E by crossing a red edge, the red T by crossing a blue edge, the red S by crossing a yellow edge, and the red A by crossing a green edge.

It is very baffling to display only the MG when doing the prediction tricks. Most people will not be able to discern just how the tricks are done by merely looking at the diagram. Suppose, for instance, Mark says yes to the red, blue, and green questions. On the MG we start at the blue D and cross the red, blue, and green edges (in any order) and will arrive at the blue S. The reader may verify that these yeses could only occur for Mark convivially choosing S.

Dropping the green question means that we ignore the red letters and ignore the green edges. The MG is then used as before with the other three colors to identify Mark's choice. A careful study of the MG will make clear how it can be used in each of the other dimensional cases.

"Impossible" Foldings

Luc De Smet

My interest in folding paper and especially playing cards started when I saw some "impossible" playing cards made by Angus Lavery. While trying to find out how they were folded I found the following basic procedures:

First Impossible Folding

Starting from	End result

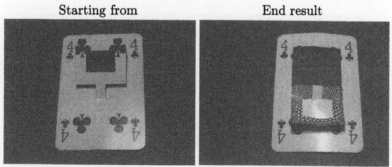

(See Color Plate VIII for a full color plate of this illustration.)

Note how the square frame is folded down, but the leg (or spine) is on the *wrong* side of the frame. One would expect the leg to be *under* the frame rather than *over* it. I'll present two ways to accomplish this feet:

Luc De Smet is a puzzle collector who is especially interested in impossible objects.

First technique: Your goal is to turn the card inside-outside, passing the outside frame of the card through the inside frame. Begin by pushing the top of the outside frame through the inside (smaller) frame from behind toward the front. (See picture above.) Continue to draw the lower (uncut) part of the card through the inside frame until the whole outside has passed through. For stiff playing cards, you'll probably have to crease the outer frame as you work the card.

Second technique: This technique is more appealing for stiff material like playing cards, because the smaller, inner frame is creased rather than the outside frame. This time it's the inner frame that is turned inside-out around the leg, and the outer frame is left alone.

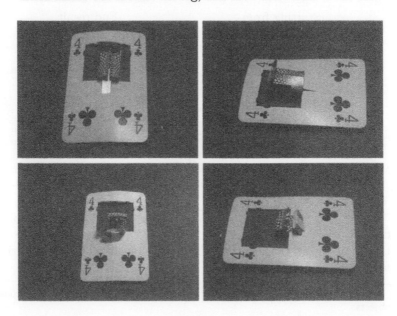

Step 1: Fold the four inner frame sides (left, right, upper, and lower in any order) inwards and toward the front as in the figure on the previous page. Note that the lower side tucks behind the central leg which is attached to the rest of the card.

Step 2: Continue to turn the inner fame inside out by passing the folded left, right and top sides of the frame down and through the middle of the frame. It's important to keep the lower side of the frame tucked behind the leg as you work. Unfold the inner frame and then fold the three sides unattached to the leg downwards onto the lower part of the card to complete the impossible fold.

 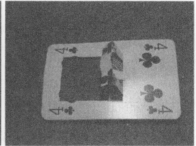

Second Impossible Folding

Starting from the same position, the end result is:

 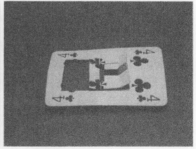

In this folding, the square frame is shifted downwards, and the leg appears mysteriously on the *top* side of the frame. Again, we can accomplish this folding in one of two ways which closely parallel the first folding.

First technique. Push the top of the outside frame through the inside (smaller) frame from the front to behind, and continue to draw the lower part of the card through the inner frame until the whole of the outside has passed through.

Second technique. For stiff material such as playing cards, I prefer:

Step 1. Fold the four frame sides of the inner square, this time toward the back of the card. The lower side should remain in front of the leg.

Step 2. Continue to pass the inner frame through itself and unfold the four sides of the frame. Note that the lower side comes under the central leg.

The second folding is closely related to the first: If you do the first folding with the back of the card facing you, then carefully move the inner frame to the front of the card, you'll arrive at this folding.

Further Combinations

Once you've mastered the basic procedures you can try a lot of combinations and variations:

- Change the shape of the frame to a rectangle, trapezium, etc....

- Don't cut a hole but keep one side of the inner part attached to give the card a "window."

- Put 2 cards one through the other.

- Make combinations of, say, 2 frames as below.

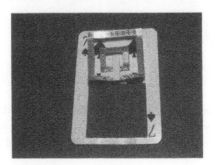

Bibliography

[Mor88] Scot Morris. *The Next Book of Omni Games*. New American Library, New York, 1988. The Museum of the Impossible.

[Rog98] Terri Rogers. *Buckled Bunkum in Top Secrets*. Breese Books, London, 1998.

Part IV
Chef's Caprice

Coincidences

A. Ross Eckler

What is a coincidence? The best definition I have seen was offered by
Frederick Mosteller and Persi Diaconis [MD89]:

> A coincidence is a surprising concurrence of events per-
> ceived as meaningfully related, with no apparent causal con-
> nection.

The key words are "concurrence of events" which can mean many
different things, from actual meeting of people in unexpected places to
similarity of characteristics of two different individuals. Although most
coincidences seem to involve people, there can also be coincidences involving abstract entities, such as the occurrence of an unusual word in
one's reading at the same time as it is heard on radio or TV. Another
example is provided by Martin Gardner's *The Wreck of the Titanic Foretold?* [Gar86b] which discusses various parallels between the real Titanic
and a fictional steamship Titan, described in the novel *Futility* written
fourteen years before the Titanic struck the iceberg.

Although the Mosteller–Diaconis definition of a coincidence is comprehensive it does not include all events which are often labeled as coincidences in popular writing. One example is the unexpectedly quick
conclusion to a search for some unknown fact or object. This is illustrated by an event which my daughter Lois experienced many years
ago. When visiting the National Cathedral in Washington, D.C., she
wanted to locate the chair given in memory of her great-grandmother.
Not knowing where it was, she picked an aisle at random—and found
the chair in that aisle! Surprising as this was, it was not a concurrence
of events in the Mosteller–Diaconis sense.

For the past 30 years **A. Ross Eckler** has edited *Word Ways*, the only journal
in the world devoted to wordplay.

Coincidences resist classification. To show how protean coincidences can be, consider the one that Martin Gardner himself has called "far and away the most startling of literary coincidences" [Gar97]. Psalm 46 of the King James Bible contains the word SHAKE as the 46th word from the start, and SPEAR as the 46th word from the end, not counting the cadential SELAH. (Recall that Shakespeare was 46 years old in 1610, the year the King James version was issued.) The "explanation" for this? It has been proposed that this was Shakespeare's literary signature, to prove that he was one of the committee of scholars revising the Bible. Unfortunately, there is no historical evidence to support such a claim, and in fact this curiosity appears in earlier versions of the Bible!

Coincidences exist to be disproved—and, therefore, to be eliminated as coincidences. The rationalist adopts the premise that all coincidences are explainable if only one knows enough about the attendant circumstances. To know enough, unfortunately, is not limited to an awareness of the details of the event itself (for example, the travel habits of two people that meet in a strange place) but requires one to assess the class of all similar events that would have been called coincidences had they been observed. Should one not consider the chance that any two persons meet in an odd place, not just the two under scrutiny? This is, in fact, akin to the well-known birthday problem, in which the chances are about 50-50 that two people in a group of 23 will have a birthday the same day of the year.

Unfortunately, most coincidences are harder to quantify than the birthday one. Mosteller and Diaconis present various simple generalizations, such as more than two people with matching birthdays, or two people with birthdays less than k days apart. An open-ended version that they did not consider assesses the likelihood that two famous people share the same day, month, and year birthday. The most noted example of this is Abraham Lincoln and Charles Darwin, both born on February 12, 1809. Should we be surprised by this? Not if there is a 50-50 chance that two people of similar fame share a birthday. Mosteller and Diaconis suggest the approximation $N = 1.2c^{1/2}$ to yield a 50-50 chance that two of N individuals will match one of c different birth dates. What value of c should be chosen? Presumably one is interested in famous people of any era, even as far back as the Roman empire, but usually one does not know exact birth dates for ancient notables. Let us see how the necessary number of people increases with the time-interval under consideration:

> 100 years 229 people
> 500 years 512 people
> 1000 years 724 people

During the past 500 years, are there 512 people as famous as Lincoln and Darwin? Over a thousand-year period, are there 724? The answer is probably yes, even though one person's most-notable list would differ from another's. In other words, the Lincoln-Darwin coincidence is to be expected, and not to be regarded as one.

This coincidence immediately suggests another, the well-known fact that Thomas Jefferson and John Adams, signers of the Declaration of Independence, both died July 4, 1826, the fiftieth anniversary of that event. How likely is this coincidence? Start with the assumption that one would have been equally impressed if any two of the 56 signers had died on the day in question, not just Jefferson and Adams. (The fact that both enjoyed more famous careers than the others, ending up as President, should not be factored in. In fact, it can be argued that their fame might have contributed to a reduction in the surprisingness of the event, as will be discussed later.) The name of each signer, together with his year of birth and exact date of death, if known, is given in Table 1.

John Adams	1735–Jul 4 1826	Thomas Lynch Jr.	1749–end of 1779
Samuel Adams	1722–Oct 2 1803	Thomas McKean	1734–May 8 1806
Josiah Bartlett	1729–May 19 1795	Arthur Middleton	1742–Jan 1 1787
Carter Braxton	1736–Oct 10 1797	Lewis Morris	1726–Jan 22 1798
Charles Carroll	1737–Nov 14 1832	Robert Morris	1734–May 8 1806
Samuel Chase	1741–Jun 19 1811	John Morton	1724–Apr 1777
Abraham Clark	1726–Feb 15 1794	Thomas Nelson Jr.	1738–Jan 4 1789
George Clymer	1739–Jan 23 1813	William Paca	1740–1799
William Ellery	1727–Feb 15 1820	Robert Treat Paine	1731–May 11 1814
William Floyd	1734–Aug 4 1821	John Penn	1741–Sep 1788
Benjamin Franklin	1706–Spr 17 1790	George Read	1733–Sep 21 1798
Elbridge Gerry	1744–Nov 23 1814	Caesar Rodney	1728–Jun 29 1784
Button Gwinnett	1732–May 27 1777	George Ross	1730–Jul 1779
Lyman Hall	1724–Oct 19 1790	Benjamin Rush	1745–Apr 19 1813
John Hancock	1737–Oct 8 1793	Edward Rutledge	1749–Jan 23 1800
Benjamin Harrison	1740–Apr 1791	Roger Sherman	1721–Jul 23 1793
John Hart	1711?–1780	James Smith	1720–Jul 11 1806
Joseph Hewes	1730–Nov 10 1779	Richard Stockton	1730–Feb 28 1781
Thomas Heyward Jr.	1746–Mar 6 1809	Thomas Stone	1743–Oct 5 1787
William Hooper	1742–1790	George Taylor	1716–Feb 23 1781
Stephen Hopkins	1707–Jul 13 1785	Matthew Thornton	1714–Jun 24 1803
Francis Hopkinson	1737–May 7 1791	George Walton	1740–Feb 2 1804
Samuel Huntington	1731–Jan 5 1796	William Whipple	1730–Nov 28 1795
Thomas Jefferson	1743–Jul 4 1826	William Williams	1731–Aug 2 1811
Francis Lightfoot Lee	1734–Apr 3 1797	James Wilson	1742–Aug 21 1798
Richard Henry Lee	1732–Jun 19 1794	John Witherspoon	1723–Nov 15 1794
Frances Lewis	1713–Dec 30 1802	Oliver Wolcott	1726–Dec 1 1797
Philip Livingston	1716–Jun 12 1788	George Wythe	1726–Jun 8 1806

Table 1. Signers of the Declaration of Independence, year of birth, and date of death (if known).

The probability that two signers died July 4, 1826 can be calculated using life tables. Reliable life tables for the eighteenth century are not available, so modern ones were used instead; however, life expectancies of men 50 to 60 years of age (when many signed the Declaration) haven't changed a great deal since that time. The calculation is straightforward but tedious. For example, for the two youngest signers, Edward Rutledge and Thomas Lynch, both 27 in 1776:

$$\mathbf{Pr}\,\{\text{dies Jul 4 1826}\} = \mathbf{Pr}\,\{\text{lives from 1776 to Jan 1 1826}\}\,\mathbf{Pr}\,\{\text{dies Jul 4}\}$$
$$= \frac{27858}{90796} \cdot \frac{2735}{27858} \cdot \frac{1}{365} \approx 0.000083$$

Similar probabilities can be calculated for the other 54 signers. The probability that some pair of signers dies on July 4 is equal to the sum of the products of these probabilities, taken over all 1540 possible pairs, a total of 0.0000007. Surely one's surprise is well-founded?

But this anniversary is one of a set of similar anniversaries. Wouldn't two signers dying on any July 4 be noteworthy? This could reduce the probability by a factor of at least 50. More interestingly, what about the possibility that people near death can, by an act of will, postpone this event a short while, say a month or so? There would have been a powerful motivation for Jefferson and Adams to survive until the 4th, particularly since both were such famous men. (Adams at least was aware of the significance, for he reportedly said on the 4th "Jefferson still lives" although in fact Jefferson had died a few hours earlier.) If both men could delay death up to a month, in effect selecting the date they would die on, then the probability cited above is increased by a factor of 1000. Under the less-likely event that death could be postponed a full year, an additional multiplicative factor of 100 is created, leading to a final probability of 0.07 for the July 4 event. Although this coincidence cannot be disposed of as neatly as the Lincoln–Darwin one, two partial explanations have been given, and others may yet be discovered.

Paul Kammerer, a German biologist who committed suicide in 1926, perhaps as the result of allegations that he had faked experimental data, was perhaps the first person to make a systematic study of coincidences. For some fifteen years he collected examples of coincidences, both from newspapers and from personal experience. One hundred are briefly described in his book *Das Gesetz der Serie* [Kam19]. Most of these contain too few details to permit analysis of the likelihood of similar coincidences. He also set up an elaborate but murky taxonomy of coincidence, making distinctions more apparent to him than to the reader. A sample conclusion:

We thus arrive at the image of a world-mosaic or cosmic kaleidescope, which, in spite of constant shufflings and rearrangements, also takes care of bringing like and like together.

One of Kammerer's examples is cited because it exemplifies a common class of coincidences:

In the Kattowicz hospital in 1915 there lay two soldiers, both 19 years old, both sick from pneumonia, both from Silesia, both volunteers in the Transport Corps, and both named Franz Richler. The one lay near death; when the relatives of the other were mistakenly notified [of his incipient demise], they found the physical similarity of the namesake so great, that they could not regard the Richler lying there in agony a stranger...

It seemed like an extraordinary coincidence to the families involved—but it is possible that it is no more amazing than winning the lottery. Such a coincidence may be nearly inevitable when one considers the population of potentially-similar events: any two soldiers in a German hospital at any time during the war, having the same name plus three or four other characteristics in common. How many characteristics could have been matched? It is likely that patient dossiers included many facts besides age, branch of service and place of origin—possibly date of enlistment, date of admission to hospital, identification number, rank, names of parents, and so on. In the absence of comprehensive statistics on German hospital patients, it is impossible to quantify the correlations that likely exist among different attributes. To begin with, it seems likely that most of the soldiers in the hospital came from Silesia, for Kattowicz was then on the Silesia–Russia frontier. Were patients sorted out, diseased patients going to one hospital and wounded ones to another? How common was the surname Richler in Silesia? Is it possible that soldiers in the same outfit enlisted together from the same locality?

Mosteller and Diaconis proposed a simple approximation for such comparisons. If there are k different characteristics with equally likely alternatives numbering $c_1, c_2, \ldots, c_k$, the number of people N required to have a 50-50 chance of a match in one characteristic is:

$$N \approx 1.2 \sqrt{\frac{1}{\frac{1}{c_1} + \frac{1}{c_2} + \cdots + \frac{1}{c_k}}}$$

What is needed is a generalization of this formula giving the population size needed for a 50-50 chance of matching any m out of a population of k different characteristics. With such a formula, one could explore how the likelihood of coincidences like Kammerer's vary as a function of the c_i, m, and k.

If k is much larger than m, it becomes ridiculously easy to find coincidences, as exemplified by various frequently-cited Lincoln-Kennedy lists. Mindful of this, the Spring 1992 issue of *The Skeptical Inquirer* [Ske] announced a "Spooky Presidential Contest." The Winter 1993 issue reported that one of the two cowinners discovered 16 parallels between John Fitzgerald Kennedy and Mexican president Alvaro Obregon. The other winner found at least six coincidences between 21 different pairs of US presidents; for Thomas Jefferson and Andrew Jackson he found thirteen.

If one cannot explain a coincidence, the temptation is to postulate some otherwise-undetectable mechanism like C. G. Jung's synchronicity (see [Jun73]). In the more specialized field of genealogical research, Henry Z. Jones recently authored two books [Jon93] and [Jon97], telling how various researchers serendipitously found various clues concerning their ancestry. Jones argues that such finds demonstrate the desire of long-dead ancestors to get in touch with their descendants, a feat accomplished by tapping into Jung's "collective unconscious." Kammerer also sought general explanations, but he relied more on analogies with physical principles such as gravity or magnetism than on parapsychology.

As a corrective for the all-too-common tendency to look for a universal explanation, I propose a simple project: a personal diary of coincidences. This will demonstrate to you how frequently coincidences occur; most coincidences we don't even notice because we are not on the lookout for them. You will see that coincidences run the gamut from ho-hum to how-amazing, with all intermediate cases represented. Confronted with a spectrum of coincidences, one is less likely to look for supernatural explanations for the more surprising ones. Do not be surprised if your coincidences, by and large, are far more mundane than ones reported in newspapers; after all, the latter represent the most startling coincidences experienced by a large population—by those who, so to speak, have won the coincidence lottery.

It is advisable to collect only coincidences which can be documented. This eliminates such coincidences as dreams that apparently foretell events, or coincidences involving a telephone call from a particular person shortly after you had been thinking of him. Such coincidences, relying on the veracity of the teller, are prone to abuse by unscrupulous individuals.

It goes without saying that you should record as much corollary information as possible relating to the coincidence. Such information may help in calculating probabilities, or at the very least may show that a coincidence is much less surprising than it appeared at the outset.

As you collect coincidences, you may be tempted to classify them as Kammerer did. My collection, numbering several dozen over a thirty-year period, fall into the following categories, many of which have already been alluded to. (You may, of course, devise others.)

- Strange Encounters: meeting a friend or acquaintance in a place remote from your (or his) usual environs.

- Our Mutual Friend: two of your friends in different areas of your life know each other, or share some unusual attribute.

- Doppelgänger: another person shares several attributes with you.

- Twice-told Tales: you encounter an unusual word or name from two independent sources in a short period of time.

The third and fourth are specifically modeled in the Mosteller–Diaconis paper. The first two, although more commonly encountered in everyday life, are extremely hard to represent with an idealized mathematical model. Such a model will inevitably incorporate many parameters for which the values (or even a probabilistic range of values) are unknown. It may well be that, for some values of the parameters, the probability of the event is fairly large (say, greater than the 0.05 or 0.01 level beloved by statisticians). Should one as a result conclude that the event has been "explained"? Should one continue to construct models until one is found for which the probability is greater than 0.01 or 0.05, for most of the "reasonable" parameter values?

There is a possible way out of this dilemma. The Central Limit Theorem of statistics states that the sum of a large number of independent random variables (for example, the total number of heads in a large number of coin-tosses) can be closely approximated by a bell-shaped curve known as the Gaussian (normal) probability distribution function. Suppose that the "surprise" of each of many similar events can be quantified—expressed by a real number. If this collection of numbers can be fitted to the (right-hand) tail of a Gaussian probability distribution function, one can plausibly argue that these events are individually created by the cumulation of a large number of sub-events. (In the case of Strange Encounters, these sub-events would be the many decisions of each individual contributing to his arrival at the place in question at the right time.) There is no attempt to separate surprises into two groups according to whether or not they can be explained by the model being used; the model says only that some events are more surprising than others. The reason for different levels of surprise cannot be explained except in general terms.

Bibliography

[Gar86b] Martin Gardner, editor. *The Wreck of the Titanic Foretold?* Prometheus Books, Buffalo, New York, 1986.

[Gar97]   Martin Gardner, Dec 1997. Personal communication.

[Jon93]   Henry Z. Jones. *Psychic Roots: Serendipity & Intuition in Genealogy.* Genealogical Publications, 1993.

[Jon97]   Henry Z. Jones. *More Psychic Roots: Further Adventures in Serendipity & Intuition in Genealogy.* Genealogical Publications, 1997.

[Jun73]   C. G. Jung. *Synchronicity: An Acausal Connecting Principle.* Princenton University Press, Princeton, N.J., 1973. Translated by R. F. C. Hull.

[Kam19]   Paul Kammerer. *Das Gesetz der Serie.* Deutsche Verlags-Anstalt, Stuttgart/Berlin, 1919.

[MD89]   Frederick Mosteller and Persi Diaconis. Methods for studying coincidences. *Journal of the American Statistical Association,* 84(408):853–861, December 1989.

[Ske]   *The Skeptical Inquirer.* The quarterly journal of the Committee for the Scientific Investigation of Claims of the Paranormal, 1978–.

The Bard and the Bible

Allan Slaight

Some years ago, I acquired Martin Gardner's 1985 Prometheus publication *The Magic Numbers of Dr. Matrix*, which contains Gardner's *Scientific American* columns on this incredible numerologist spanning a period from 1960 to 1980. There I encountered the intelligence that in the 46th Psalm in the King James version of the Bible, the 46th word is SHAKE and the 46th word counting from the end is SPEAR. Dr. Matrix espoused the incredible theory that William Shakespeare worked secretly on part of the King James translation of the Bible. Matrix fortified his position by revealing that Shakespeare was 46 when the King James version was completed in 1610, and that he was born on April 23 and died on April 23 and that twice 23 is 46. This information, coupled with a careful reading of *Shakespeare* (1956) by F. E. Halliday, led to the following unusual demonstration.

"Critics and scholars have argued for years over whether the Bible or the work of Shakespeare contains the finest writing in the English language. These comparisons are invariably made using the King James version, a copy of which I have here. Let me put the debate to rest by proving to you that William Shakespeare himself actually was the anonymous author who translated and rewrote the King James version of the Bible. Shakespeare hid a series of telling clues while preparing his translation and I will attempt to reveal then through numerology.

"Naturally we will refer to the Book of Numbers, and we will make a square of sixteen numbers. If we read in order from the Table of Contents we have Genesis, Exodus, Leviticus, Numbers. As Numbers is the fourth Book in the Bible, we will start our numbering with "4".

Allan Slaight was the co-editor, with the late Howard Lyons, of *Stewart James In Print* (1989). He subsequently authored *The James File* (2000), a two-volume set totaling some 1700 pages.

"And I will have you choose four numbers freely. Please circle any of the sixteen numbers, then draw a line through the other numbers in that row and that column. Now circle any number which has not been eliminated and draw a line through that row and column. Do this a third time. There is one number left, so please circle it."

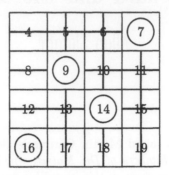

In the above example, the spectator first circled "9" and then drew lines through the other numbers in that row (8, 10 and 11) and the other numbers in that column (5, 13 and 17). He then circled "14" and did the same thing, then "7" and repeated the elimination of the other numbers in that row and column (4, 5, 6 and 11, 15 and 19). Only "16" was left, and that was the fourth number he circled.

The spectator is then instructed to total the four circled numbers. The result is always 46, regardless of numbers chosen via this process.

"Your freely chosen number is 46, but you will soon realize that—in numerology—no choice is totally free and without consequence. Please open the Bible to the first chapter of Numbers, and find the 46th verse. Read it to us."

"Even all they that were numbered were six hundred thousand and three thousand and five hundred and fifty."

"Let's jot those numbers down and add up their digits."

$$
\begin{array}{r}
600,000 \\
3,000 \\
\underline{550} \\
603550
\end{array}
\quad \text{and } 6 + 0 + 3 + 5 + 5 + 0 = 19
$$

"Isn't it interesting that the last number in our square is '19'? Now please refer to the Table of Contents again, and name the Book in nineteenth position. Psalms? Please open the bible to Psalms and use your freely selected number ... 46. Find the 46th Psalm and count to the 46th word. What is the word? SHAKE. Now count to the 46th word from the end and call out that word. SPEAR. SHAKE-SPEAR.

"Through numerology, you have discovered one of the scores of clues Shakespeare cunningly concealed when he was translating what became the King James version. However, there are skeptics today who still maintain that others wrote the plays attributed to Shakespeare. Bacon or Marlowe, for example. There is one way of determining if this theory is accurate or false, and that is by using the other numbers in our square of sixteen. Please total the twelve numbers you did not circle. Your answer is 138?

"Please look at the 138th Psalm, but I won't ask you to count to the 138th word. Just total the three digits in 138: $1 + 3 + 8 = 12$. Count to the twelfth word in the 138th Psalm. What is the word? WILL! WILL SHAKE-SPEAR. Proof positive.

"Should you wish for more evidence through numerology, let me mention that William Shakespeare was born an April 23 and died on April 23. April is the fourth month, just as Numbers is the fourth Book. And the day of birth and death, both 23, total 46 which is the numerical key you discovered. And it should be mentioned that 23 is the number of the best-known Psalm.

"In addition, the first of Shakespeare's plays to be published with his name credited was *Love's Labour Lost* in 1598 and those digits total 23! And the folio of all his plays was first issued in 16<u>23</u>! Of course, those two 23s total 46. Another point: the King James version was completed in 1610—when Shakespeare was 46 years old!

"And finally, if we assign numbers to the letters in 'Shakespeare' like this:

1	2	3	4	5	6	7	8	9	0
S	H	A	K	E	S	P	E	A	R
E									

you will find they total...46: It *is* rather amazing that you also chose the number 46."

Comments and Credits

My first objective was to uncover a technique to force the number "46" in a manner compatible with the rest of the scenario. The matrix concept to force "34" with numbers from 1 to 16 led rather easily to the "46" outcome with numbers from 4 to 19. The principle seems to have first appeared in print in Maurice Kraitchik's *Mathematical Recreations* [Kra42] although we should credit Walter Gibson with the seminal idea in his Date Sense [Gib38]. Mel Stover, Stewart James, Martin Gardner, Howard Lyonsand Sam Dalal have produced significant variations, and likely the finest presentation employing this principle is Phil Goldstein's Rainbow Matrix [Gol90].

I'll confess that unearthing the number "19" in the 46th verse of the first chapter of Numbers to lead the spectator to the book of Psalms generated a momentary thrill, but that was small potatoes when compared to the discovery that the eliminated numbers in the matrix always totalled 138 and logically produced the word "will". Of course, Dr. Matrix would have stated it was inevitable that assigning numbers, as described, to the letters in "Shakespeare" would produce that particular total.

Bibliography

[Gib38] Walter Gibson. Date sense. *Jinx*, 41, February 1938.

[Gol90] Phil Goldstein. *Violet Book of Mentalism.* (privately published), 1990.

[Kra42] Maurice Kraitchik. *Mathematical Recreations.* W. W. Norton, New York, 1942.

Inversions: Lettering with a Mathematical Twist

Scott Kim

Graphic Palindromes

"Inversions" is my name for words written so they can be read in more than one way. For instance, the word INVERSIONS below becomes my name when you turn it upside down.

Inversions

As a child I was fond of mathematics, magic, music and wordplay. I was aware of palindromes—words like "racecar" that are spelled the same backwards and forwards—but only knew of a few words like NOON that actually look the same right side up and upside down. It wasn't till I took my first graphic design class in college that I created my first inversion. Suddenly there was a whole new world to explore.

What began as a hobby has blossomed into a personal art form that I continue to develop many years later. In 1981 I collected sixty of my lettering designs into a book called *Inversions* [Kim81] Other people who create inversions include cognitive scientist Douglas Hofstadter, who coined the term ambigrams as the generic term for the art, John

Scott Kim (www.scottkim.com) is an independent puzzle designer. Work includes *Inversions* (book), Railroad Rush Hour (toy), *Discover* magazine "Boggler" (monthly puzzle), and *The Next Tetris* (computer game).

Langdon, who wrote his own book of symmetrical lettering designs called *Wordplay* [Lan92], and calligrapher Lefty Fontenrose.

In honor of Martin Gardner, I would like to share some of my more mathematically minded inversions. These designs are mathematical in three senses: they use mathematical symmetries, the words and names are drawn from the world of mathematics, and the process of creating an inversion is akin to solving a mathematical problem. Elsewhere in this book you will also find inversion tributes to mathematician David Klarner, educator Harry Eng, and magician Mel Stover.

Martin Gardner

I first met Martin Gardner at his home in New York in 1976 when I was flying around the country looking at graduate schools. I brought with me a folder of inversions to show him. Never at a loss for an interesting comeback, he responded by showing me that the value of the mathematical constant pi written out to two decimal places says something interesting in a mirror. Hold this page up to a mirror and see what happens!

3.14

This inversion on MARTIN GARDNER appeared in *Inversions*.

After *Inversions* appeared, Peter Renz of W. H. Freeman and Company asked me to do chapter opening illustrations for Gardner's book *Aha! Gotcha* [Gar82]. This illustration opened the chapter on Numbers. Every numeral appears at least once in the names of the numerals.

0	ZERO	5	FIVE
1	ONE	6	SIX
2	TWO	7	SEVEN
3	THREE	8	EIGHT
4	FOUR	9	NINE

In 1993 puzzle aficionado Tom Rodgers started the Gathering for Gardner. I decided to revisit my original MARTIN GARDNER inversion, to see what else I could do. First I tried a different symmetry. Hold a mirror against the page in a horizontal line just below this word, and you will be able to read both the first and last names.

Then I tried a different combination of words. This time MARTIN GARDNER turns not into itself, but the name of his alter ego DOCTOR MATRIX, who appeared many times in his *Scientific American* column.

Finally I paid tribute to his *Scientific American* column, noting that the column and the person both had the same initials. Read the small letters of the first word and the big letters of the second word to spell the name of the column; read the big letters of the first word and the small letters of the second word to spell the name of the person.

Mathematics

Some of the first words I tried as inversions come from the world of mathematics. MATHEMATICS is the same when rotated 180 degrees. UPSIDE DOWN is also the same upside down. Notice that a six-letter word becomes a four-letter word.

MIRROR is the same in a mirror, but not upside down. People often try to turn this design upside down, confusing rotational and reflective symmetry.

TRUE/FALSE is not symmetrical, but does express a duality. IN-FINITY is not only infinite, it reads both clockwise and anti-clockwise. Notice that FI becomes Y.

TREE. Notice that each E is also a T that begins a smaller TREE.

Education

Although I am not a classroom teacher, I have a strong interest in education. Here are three poster images I have created for teachers.

INVERSIONS NAMES (Dale Seymour Publications). Kids like to see inversions of their own names or names of friends. In this poster there are 26 first names, one for each letter of the alphabet. Each name is exactly symmetrical. Different names have different symmetries. Each lettering style appears twice, once as a boy's name and once as a girl's name. Names that start with letters at opposite ends of the alphabet— e.g., A and Z, B and Y—appear on opposite sides of the design.

ALPHABET SYMMETRY (Dale Seymour Publications). The entire alphabet, in exact mirror symmetry; originally produced for the Museum of Fun in 1985, an exhibit of illusionary art that toured Japan.

TEACH/LEARN. Originally designed for the Apple Multimedia Laboratory's pioneering interactive video disc The Visual Almanac. It expresses the dual nature of teaching: when you teach you learn.

Mathematicians

When I design an inversion I am always interested in expressing the meaning of the word or name. PYTHAGORAS is full of triangles.

MOEBIUS uses glide reflection symmetry, which allows it to connect with itself seamlessly when written on a transparent Moebius strip.

Every letter in JOHN CONWAY is an oscillator (a pattern that eventually returns to its original state) from Conway's game of Life.

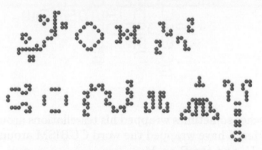

I have drawn graph theorist Frank HARARY's name as a graph.

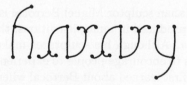

Solomon GOLOMB invented pentominoes(r): the twelve shapes that can be formed out of five unit squares.

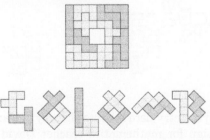

Art and Mathematics

M. C. ESCHER. Escher elegantly combined art and mathematics in his work to express cosmic ideas. This pattern was inspired by Escher's many periodic tessellation drawings. Escher occasionally dabbled in visual wordplay, such as the beginning of the scroll Metamorphose.

CUBISM. Escher sometimes wrapped his tessellations around the surface of a sphere. Here I have wrapped the word CUBISM around a cube. The C becomes a U, and the B an M.

BERROCAL. The Italian sculptor Miguel Berrocal is best known for his tabletop sculptures, produced in multiple editions, that come apart into as many as 90 pieces. Although his sculptures make excellent puzzles, his real motivation is to encourage people to experience sculptural forms by handling them. I first learned about Berrocal when I visited Gardner.

ORIGAMI. The symmetry here mirrors the way origami paper is folded. I created this design for mathematics teacher David Masunaga of the

Iolani School, Honolulu, Hawaii, who caught the art/math bug at Arthur Loeb's Design Science program at Harvard University. It first appeared in print in *Origami A to Z* by Peter Suber.

Puzzle People

SLOCUM. Every few years I meet my puzzle-loving colleagues from around the world at the International Puzzle Party, founded by puzzle collector and historian Jerry Slocum of Beverly Hills, California. For a man so identified with puzzles I wanted to find PUZZLES in his name.

EDWARD HORDERN. Edward Hordern of England was the second pillar of the International Puzzle Party. His specialty is solving puzzles. I often refer to his book of sliding block puzzles and their solutions. By rotating and sliding the elements of his first name, you can make his last name.

NOB YOSHIGAHARA. The third pillar of the International Puzzle Party is NOB Yoshigahara of Tokyo, Japan, a prolific puzzle inventor and author. I greatly admire his creations, such as the sliding block puzzle Rush Hour and his more recent Lunar Lockout, both manufactured by Binary Arts. NOB uses this profile as a symbol; here I found a way to rotate pieces of his profile to make his given name. (The extra lines in his head show the outlines of the letters, and are not part of his signature profile.)

NOB was disappointed that I couldn't do my standard rotationally symmetric inversion on his name, so I kept trying, and almost 20 years after my first attempt I found this solution.

Magicians

The Gathering for Gardner attracts some of the world's leading magicians, reflecting Gardner's interest in magic.

MAX MAVEN. When I first met mentalist Max Maven at the Magic Castle in Hollywood, I did a conventional inversion on his name. But that seemed too bland for such a sharp character. Using an old magic principle, the design below is meant to be shown with your thumb covering the E. Move your thumb and reveal that what appears to be his first name is really his last name.

MEIR YEDID is best known for Finger Fantasy, an act in which he makes fingers disappear from his hand. I was pleased to find a way to work his first name, last name, and act into a single design.

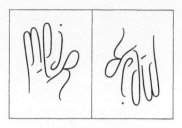

Bibliography

[Gar82] Martin Gardner. *Aha! Gotcha: Paradoxes to Puzzle and Delight.* W. H. Freeman, San Francisco, 1982.

[Kim81] Scott Kim. *Inversions: A Catalog of Calligraphic Cartwheels.* BYTE Books, Petersborough, N.H., 1981.

[Lan92] John Langdon. *Wordplay: Ambigrams and Reflections on the Art of Ambigrams.* Harcourt Brace Jovanovich, New York, 1992.

All You Need Is Cards

Brain Epstein

What follows is a peek into the lives of four lads during one of their most creative periods. Several effects with cards are considered, peppered with fantastic and fabulous references. All will be explained in due course.

You Know My Suit (Look up the Number)

The boys were still groggy after a late night of recording when they reassembled at noon for rehearsals. John was the last to arrive, greeting his three friends with as enthusiastic a "Good morning, good morning," as he could muster. George was squatting on the floor, reading *Quartet*, a collection of short stories by the paperback writer Somerset Maugham, which Paul had given him. Nobody was in a hurry to get to work, and there were no objections when George pulled out a pack of cards.

"John," he began, throwing clubs aside. "Guess who has real magical powers? It's not just his lucky rings! Yes, it's our little Richard!"

"A-wop bop a-loo bop a-lop bam boom! I thought that was me," Paul said, clearly a bit miffed.

"I'm so tired," said John, trying to look interested. "I haven't slept a wink. But hey, hey, hey, hey, I don't want to spoil the party. So you're on a roll, Ringo? Okay, give us your best trick then, for Pete's sake." John, Paul and George were now seated in a circle on the floor.

George handed John ten clubs in order, from Ace to Ten, as Ringo and Paul ambled off for a quick cup of tea. "Ringo!" George called after them, "Don't be long; I need you." He turned back to John. "Mix these

---

Brain Epstein learnt everything he knows about card tricks and the best ale from Mullah McCoy (a tome tempt card fetish man), Spelman College, Atlanta.

up as much as you want while Ringo is gone; then give them back to me."

"Hey bulldog!" said John, as he looked through the cards, "We've certainly played our share of clubs over the years!" When he had thoroughly jumbled the cards he handed them back to George, who looked through them briefly, and dealt them out face up, in the order in which John had handed them to him. He then silently turned four of them face down.

"Mr. H performs his tricks without a sound," observed John. "Mr. H will demonstrate," George chimed in with a grin, "Four Somersets he'll undertake on solid ground." He picked up his book again and studied it intently until Paul and Ringo returned.

The cards were lined up in this order: a face-down card, followed by the 10, 5, 6, 8, another face-down, the 2, two face-down cards, and the 4. "You know what to do," said George. Ringo surveyed the cards silently, his lips moving noticeably as he tried to concentrate.

"I know what they are!" said John excitedly. The others looked up. "Number 9, Number 9, Number 9, Number 9..." he said slowly, in a monotonous voice. Everybody groaned. Then Ringo finally spoke: "Boys, the hidden cards are the Ace, 9, 7 and 3, in that order."

"What did I tell you? I was right about one of them!" shouted John, leaping up with a laugh. He turned over the four cards, gasping when he saw that Ringo had got them all right. "By jingo, Ringo, it seems that you have the touch, all right," he said, obviously quite impressed.

"With a little luck..." said Paul under his breath. "Wait—can you do it again?" John interjected. "Not a second time. Happiness is a warm pack. I bet he can't do to it any time at all."

Paul mixed up the first ten hearts from the pack while Ringo went off to get more sugar for his tea. "Sugar daddy's lovely hearts/clubs banned," mocked Paul, as he handed the cards to George, who glanced through them, as before, and dealt them out without changing the order. However, this time he turned five of them face down with a sly grin. When Ringo returned he saw a face-down card followed by the 7, three face-down cards, the 3, Ace, another face-down card, and then the 5 and 6. "That's five you expect me to get right," he said indignantly. George smiled sweetly. "Considering how much sugar you put in your tea, I have every confidence in you."

It wasn't long before Ringo announced that the hidden cards were the 8, 4, 9, 10 and 2, in that order; once again, he was absolutely correct. John and Paul were flabbergasted, but no matter how much they begged George or Ringo, neither would utter a word of explanation.

"Ain't he sweet?" said John to the others. "Ringo, how do you do it? I want you to tell me why you did five that time, and not four."

"I've got a feeling—a feeling deep inside," said Paul. "We can work it out... what you're doing."

"It won't be long," John agreed.

"Think for yourself," was all George would say, over and over. The cards were put away, and Ringo sat magisterially behind his drum kit while the others tuned their guitars, so that they could have another go at Paul's 'Air Dish' number, which they were having a lot of trouble agreeing on.

Two days later, after an especially fruitful studio session, the lads celebrated by going out for a late night meal at an Indian restaurant called *The Inner Light* that George recommended. As soon as the plates were cleared away, John produced a pack of cards.

"Have you heard the word?" he began, with that telltale twinkle in his eye. "This boy"—indicating Paul—"has been doing a lot of yogurt lately, and as a result his mental powers are razor-sharp right now. You should have seen him last night: he was really flying." George looked sceptical.

"All I've got to do is get Ringo to help me," John said, handing the pack to the drummer. "Give me back any five cards—any five at all." A suspicious Ringo fanned through the pack, and picked out five cards, which he gave John. George watched intently from across the table. "Now, I'm going to show Paul *four* of these cards, but hide this one," said John.

He slid one of the cards over towards Ringo, who was about to pick it up when Paul admonished him. "Let it be," he said, putting the card under Ringo's packet of cigarettes. He held up the other four cards above his head for all to see. They were the Ace of Spades, Four of Hearts, Nine of Spades and Four of Diamonds.

"Help!" said John with a giggle, looking at Paul. "I need somebody—not just anybody. I do appreciate you being around; won't you please help me? What is the card beside Ringo?" The atmosphere was thick as everybody watched Paul, who screwed up his face in concentration. "Don't let me down," said John nervously. "Please please me, and get it right."

At last, Paul spoke: "Is it the Two of Spades?" Ringo pushed his matchbox and cigarettes aside, and turned the card over. "Yes it is, it's true," the other three concurred in perfect harmony.

George and Ringo were baffled. How had Paul named that card? It could have been anything! "Do you want to know a secret?" asked Paul.

"Here's another clue for you all, the adder is Paul," offered John.

"One and one is two," Paul teased, but the others were none the wiser.

A repeat performance was demanded, and Paul and John were happy to oblige. George chose five cards and handed them to John. As before, John looked at them carefully before putting one aside. He held the Queen of Clubs, King of Diamonds, Nine of Diamonds, and Five of Diamonds high above his head. *Floozy in the sky with diamonds*, he thought to himself.

"This one is harder," said Paul, "But it's on the tip of my tongue." After a pause, he inquired, "The Five of Clubs?" He was, of course, correct.

"It's all too much," commented a thoroughly confused George.

"Ask me why I held up the cards the way I did," said John. Ringo was looking at the sugar bowl longingly, so George took the bait. "Because"— John paused enigmatically—"Baby's in black, and I'm feeling blue." Everybody looked at the black queen. "Very helpful," said George sarcastically, "As if I needed someone to tell me that!"

"That means a lot," John insisted, "Would you rather I gave you no reply at all? I should have known better than to try to give you a hint."

"Every little thing is important," said Paul, trying to smooth the waves. "She's a woman." He pointed at the black Queen, and then at the other cards, adding, "Here, there and everywhere." George stared ahead blankly.

"There's a place, not far from here George," Paul confided, "Where John and I sat down yesterday, just the two of us, and worked out that one together. It's as good as the trick I remember you and Ringo doing for us the night before."

"I'm off," said a clearly exhausted Ringo, standing up and taking a pair of gloves off the table. "It's been a hard day's night."

"Leave my mittens alone," snapped John.

"Sorry," said Ringo, dropping them hastily. They walked outside.

"Misery," said John, buttoning up his coat. It was a dark and stormy night. "If the rain comes, run and hide your heads."

"Johnny, you can drive my car," offered Paul. "In spite of all the danger, George and I are going to walk."

"Though the taxman would probably like to tax my feet," George said, looking down. "These old brown shoes have seen better days."

"Run for your life," advised John, as the heavens opened.

"Now it's time to say good night," said Ringo. "Sleep tight. See you all tomorrow."

"Tomorrow never knows," said John cryptically.

"You won't see me," Paul said. "I'm meeting Dr. Robert, then I'm spending the afternoon with another girl. She says it's her birthday."

Weeks passed before the lads met again. They had been on separate holidays, but it was time to get back to work. They met for lunch on a dull grey day, eager to talk about some new material they had come up with.

"What goes on?" asked Ringo, as John and Paul came together. "I feel fine," said a particularly chipper-sounding John. "I'm down. I'm really down," moped Paul, letting his face grow long for a second before dissolving into laughter. "Would anybody like to see a card trick?" he asked a few minutes later, as soon as George had joined them. "Nobody I know," said John, who was keen to talk about several songs he had recently written. "Don't bother me," said George, who had songs of his own he wanted the others to consider. "Having been some days in preparation, a splendid time is guaranteed for Paul," joked Ringo, but nobody seemed to notice.

"Maybe we should do our new trick for no one," Paul said to Ringo, raising his eyebrows dramatically and riffling the cards noisily. "Roll up, step right this way; and that's an invitation. The magical mystery tour is waiting to take you away." John and George gave in and agreed to watch.

"With a little help from my friend," began Ringo, "I'd like to perform a new trick for you all. This one requires two people who are totally in tune to each others' every feeling, like a married couple. Paul, will you be my awfully wedded wife?"

"I will," replied Paul solemnly. He then grabbed Ringo's hand and added, "I want to hold your hand. Oh! Darling. Besame mucho."

"Honey, don't," Ringo said in mock consternation, backing off. "I wanna be your man." John and George laughed hysterically. "John, give Paul any four cards from this pack," said Ringo, as soon as he had regained his composure. Paul took the cards John offered him, glanced at them, and handed one back, saying, "From me to you." John grinned. Paul then placed the three remaining cards in a row on the table, but this time, only the middle one was face up.

"Tell me what you see," he asked Ringo. "The Ace of Hearts, in between two face-down cards," Ringo replied.

"What's this?" John asked, sliding the fourth card over next to the others. "The one after 909?"

After a short pause Ringo said, "I think it's the Three of Hearts."

"This happened once before," said John. "But you only had three cards to go by this time, not four like Paul and I did, and you can't even see two of them!" He threw the Three of Hearts on the table incredulously. "You can't do that," he said, shaking his head in disbelief, "Not every time. You really got a hold on me with this one."

"Within you, without you... is it a yin/yang thing?" asked George.

"I'd be an ass if I fell for that," John continued, ignoring him. "I detect a Spaniard in the forks." His incisive wit and fondness for spoonerisms were never far away. He turned over the two-face down cards—the King of Spades and the Eight of Clubs—and stared at them. *Everybody's got something to hide except me and my donkey*, he thought to himself.

"You've got to admit he's getting better," said Paul, who was clearly proud of Ringo's performance. "Getting better all the time."

"Couldn't get much worse," muttered John. "I'm just a jealous guy."

"Something in the way he grooves," George chipped in, "Attracts me like no other mother."

Of course they repeated the trick. George picked four cards and gave them to Paul, who handed one back to him, saying, "It's for you." Paul then placed the three remaining cards in a row on the table. This time, the first one was face down, followed by the Queen of Diamonds and the Ten of Hearts. "I've just seen a face," said Paul, pointing at the Queen. "Can't forget the time or place... she's just the girl for me." The others laughed.

"Oh, dear, what can I do?" asked Ringo, who seemed at a bit of a loss.

"Act naturally," advised Paul, tongue in cheek.

"Slow down," John suggested. "You're moving way too fast."

"What would you do if I guessed the wrong card? Would you stand up and walk out on me?" Ringo hoped he could carry that weight this time. In his nervousness, he edged closer to George and the hidden card.

"Get back," said Paul good-naturedly. Ringo retreated sheepishly.

"Come on. What card does George have?" asked John impatiently.

"I call your name," Ringo began, indicating the card with a flourish. "And you are—the Four of Spades!" He was relieved when George flipped the card over and proved him right yet again. "We hope you have enjoyed the show. All together now," said Paul, leading a brief burst of clapping.

"You're a legend in your own lunchtime!" beamed John.

It was starting to brighten up outside. "Here comes the sun; a cloudburst doesn't last all day," said George. "Good day, sunshine," Paul added with gusto. "Ah, Mr. Sun King," said John approvingly. "No more getting a tan standing in the English rain." They all felt good, in a special way.

"It took me so long to find out," Ringo mused. "But I found out!"

"When I get home I'll get it," said John. "I bet some of our earlier ideas come together in the end. I'll be back with the answer, I'll get you!"

"Remember the things we said today," said Paul. "Maybe I'm amazed at the way it fooled you, John. Another day, I'll explain it all to you."

"Love you to," George said. "Isn't it a pity it's so long, long, long?"

"It don't come easy," observed Ringo, looking at his hands. "After all that tricky card handling, I've got blisters on my fingers!"

"Imagine there's no card tricks," John shuddered. "No clubs, hearts, spades or diamonds; above us, only pi. Goo goo g'joob!"

"They've been going in and out of style, but they're guaranteed to raise a smile," Paul assured everybody.

"Across the universe, people have a real love of card tricks," said John, who always wanted to have the last word.

"All you need is cards," everybody chorused.

Come and Get It

We now set off on the long and winding road which leads to full explanations of the above. Each trick involves communication between two people using only mathematical principles; there is no physical or verbal signalling. One person assumes the role of *performer*, chosing and displaying cards carefully, while the other is a *confederate*, surveying the scene later and doing some mental calculations, before correctly identifying a hidden card or cards.

First trick. Here is a verse to ponder.

> In each list of ten there is bound,
> To be four that do rise; is that sound?
> In a paper with Erdős,
> By Gyorgy Szekeres,
> A counterexample is found.

This refers to an application—due to Erdős and Szekeres [AZ98, page 124]—of the Pigeon Hole Principle.[1] Martin Gardner explained it

---

[1] If n pigeons fly over a piece of land which is broken up into k fields, where $k < n$, and each single pigeon does what pigeons do best, i.e., leaves a deposit, then at least one field will receive two or more deposits. The same holds for blackbirds or bluebirds. It is this principle which that cold water on the whole "Eight Days A Week" idea.

in *Riddles of the Sphinx* in terms of "a row of 10 soldiers, no two of the same height... no matter what the order, there will always be at least four among the ten, not necessarily standing next to each other, who will be in ascending or in descending order" [Gar87, page 5].

Hence, if we first fix an ordering on a set of ten distinct cards, and are given these cards mixed up, then there will either be four in ascending or four in descending order. To perform the trick, you and your confederate first decide on an ordering, e.g., you could use numerical ordering from 1 to 10. The cards are jumbled. If you see four in ascending order, deal all ten cards out from left to right and then turn over these four. Your confederate can tell which four cards are turned over by observing which six are visible, and since the order of the other four is known, each can be identified correctly. If there is no ascending subsequence of length four, then there is a descending one instead. The trick can be performed as above, simply dealing out the cards in reverse order; there are many ways to to this without arousing suspicion if you haven't shown anybody the cards in advance. Often one gets lucky, and finds runs of length five or six, that's what happened the second performance of this trick. Of course, the usual ordering from 1 to 10 is rather obvious. The ordering George and Ringo used was alphabetical—Ace, Eight, Five, Four, Nine, Seven, Six, Ten, Three and Two. (A version of this trick using "Erdős numbers" (but not the same old suit) may be found in Colm Mulcahy's online AMS article "Mathematical Card Tricks," at http://www.ams.org/new-in-math/cover/mulcahy1.html.)

Second trick. The trick John and Paul performed for George and Ringo is often credited to mathematician William Fitch Cheney Jr. Martin Gardner has alluded to its 1950 appearance in a book by W. Wallace Lee [Lee60] (see *Mathematics Magic and Mystery* [Gar56, page 32] and *The Unexpected Hanging and Other Mathematical Diversions* [Gar69, page 158]).

A volunteer selects five cards from a standard 52-card pack, and hands them to you so that nobody else can see them. You glance at them briefly, and hand one card back, which is set aside. You quickly display the remaining four card faces, in a row from left to right. Your confederate merely glances at the visible card faces, and promptly names the hidden card.

In each of John and Paul's performances, note that the first of the four cards displayed was the same suit as the hidden card. Indeed, the Pigeon Hole Principle guarantees that in any set of five cards there will be at least one suit match. If there are two Spades, let's agree to use the first position of the cards held up for the retained Spade—thereby revealing the suit of the hidden card. Since there are 3! = 6 ways to arrange the other three cards, we can communicate one of six things. The trick is to do this independently of the particular cards held. Even then, there is another issue to address if we hope to pull it off every time.

To get the permutations idea to work with any three cards, just note that they are distinct! Thus, with respect to a fixed ordering of the entire pack, one of them is *low*, one is *medium*, and one is *high*. This suggests an easy way to communicate a number between 1 and 6. For instance, try this CHaSeD ordering: A♣,..., K♣, A♡, ..., K♡, A♠,..., K♠, A♢,..., K♢. Mentally label the three cards L (low), M (medium), and H (high) with respect to this ordering. Next, rank the six permutations of {L,M,H} as follows: 1 = LMH, 2 = LHM, 3 = MLH, 4 = MHL, 5 = HLM and 6 = HML. Now, order the cards from left to right according to this scheme to communicate the integer desired. Try this out to see out what integers John conveyed to Paul each time they did the trick. Are you certain that it happens all the time? After all, the hidden card could be any one of twelve Spades—try fixing a hole where the Ace gets in!

The last crucial observation we need is this: you must be careful as to exactly which Spade you hand back. Considering the 13 possible card values, 1 (Ace), 23,..., 10, J, Q, K, as being arranged clockwise on a circle, we can see that our two Spades are at most 6 values apart, i.e., counting clockwise one of them lies at most 6 vertices past the other. Give this "higher" valued Spade back to the victim, who hides it. You'll use the "lower" Spade and the other three cards to communicate the identity of this hidden card. The first time the trick was done, the first card was the Ace of Spades, the integer communicated was 1, and as Paul pointed out to the others, $1 + 1 = 2$: The hidden card was the Two of Spades. The second time, the first card was the Queen of Spades (with a numerical value of 12), the communicated integer was 6, and $12 + 6$ is equivalent to 5 (modulo 13). Sure enough, the hidden card was the Five of Spades.

One weakness in our method—especially if the trick is to be repeated— is the invariant use of the first position as the "suit-giver." Here is one way to overcome this: since both you and your confederate get to see four cards, add their values and reduce modulo 4 (using 4 if you get 0), letting the answer determine the position of the suit-giving card. Thus, a Jack, 8, 2 and 7 would result in $11 + 8 + 2 + 7 \equiv 0 \pmod 4$, so the fourth position would determine the suit-giver, and the other three cards communicate the permutation. Our method appears watertight, in that

it does not seem to extend to packs with more than 52 cards. However,
David and Hal Kierstead, Elwyn Berlekamp and others have noted that
the trick generalizes to packs of size 124 cards [KK00]. Try it for an
ordinary pack supplemented with one Lennonesque joker.[2]

Third and Fourth tricks.

The last trick performed by Paul and Ringo
is a kind of extension of the second one. Before attacking, we suggest
relabelling this as the fourth trick, and first trying the following, simpler,
"third" trick. You are given any five cards, one of which you hand back
before placing the remaining four in a row on the table, *some face up,
some face down.* Your confederate succeeds in identifying the fifth card.
Hint: George made an insightful comment here, if only somebody would
listen.

The Pigeon Hole Principle again guarantees that (at least) two of
the five cards are of the same suit, let's suppose it's Spades. You hand
one back, and use the remaining four to tell your confederate the identity of the fifth card. Use one particular position (e.g., the first) of
the four available for the retained Spade—which determines the suit of
the fifth card—and the other three positions for the remaining cards.
The difference here is that you communicate using some kind of binary
code—George *did* suggest a yin/yang principle—rather than permutations. Unlike in the earlier trick, the actual identities of any face up
cards play no role! You can communicate any one of $2^3 = 8$ integers in
this way. Is this enough?

As before, save the "lower" Spade and communicate the identity
of the "higher" one. Use a particular position (e.g., the first) for the
retained Spade. Rather than indicating actual binary representations
with the up/down arrangements, let's agree on this convention: UDD,
DUD, DDU (only 1st, 2nd or 3rd position is Up), and DUU, UDU,
UUD (only 1st, 2nd or 3rd position is Down), respectively, reveal to
your confederate which of $1, 2, 3$ or $4, 5, 6$ they should add to the lower
Spade value.

Finally, consider the last trick that Paul and Ringo did for the others.
Note that there may not be any suit matches among four cards! Start
by repartitioning the pack into three new Suits of 17 cards, leaving one
card (say A$\diamond$) aside. The new Suits are the standard suits $\clubsuit$, $\heartsuit$, $\spadesuit$,
each supplemented with four $\diamond$'s: Suit A is A$\clubsuit$, 2$\clubsuit$,..., K$\clubsuit$, 2$\diamond$, 3$\diamond$,
4$\diamond$, 5$\diamond$; Suit B is A$\heartsuit$, 2$\heartsuit$,..., K$\heartsuit$, 6$\spadesuit$, 7$\diamond$, 8$\diamond$, 9$\diamond$; and Suit C is A$\spadesuit$,
2$\spadesuit$,..., K$\spadesuit$, 10$\diamond$, J$\diamond$, Q$\diamond$, K$\diamond$. If one of the four cards is A$\diamond$, play the
others face down, and watch the audience reaction as your confederate

---

[2]Don't you think the joker laughs at you?

demonstrates some *real* magic powers! Otherwise (at least) two of the cards are from the same new Suit. Let's assume it's Suit A.

Retain the "lower" card and hide the "higher" one, whose numerical value is k past the retained card, where k is between 1 and 8 this time. As before, the displays UDD, DUD, DDU, DUU, UDU, UUD can communicate k if it's between 1 and 6. Since at least one card will be face up, we can use such a card—or the first such if there are two— to reveal the suit at the same time! However, we also need a way to communicate 7 or 8, and for this we use the UUU option. Let's agree that one particular U (say, the middle one) gives the suit. Then with respect to some ordering of the pack—such as lining up Suits A, B, C— there are two ways to play the other two: Low-High (to convey $k = 7$) or High-Low (for $k = 8$). You should now verify that Paul was indeed able to communicate the hidden card to Ringo in the last trick they did together.

Michael Trick at Carnegie Mellon kindly put together a website which illustrates this—er—*trick* in action, see: http://mat.gsia.cmu. edu/CARD/. It uses a slightly different suit convention: the three basic suits are Clubs, Diamonds and Hearts, each supplemented with four Spades.

This concludes our look at four fab four four-card tricks.[3]

Acknowledgements

Paul Zorn first alerted us to Cheney's card trick, thereby inadvertently providing the impetus for half of this document. Art Benjamin supplied the trick's source. Jeffrey Ehme, Eugene Belogay, Derek Smith, Scott Anderson and Michael Trick were willing accomplices at various times. Ann(e) Powers was our most charming and innocent confederate! Gordon Bean, Saki, Joe Gallian and Kevin Dunn provided invaluable input and perspective. Pete Winkler and Ron Gould made topological contributions. Tony Phillips gave poetic advice. Thanks to DCAM for exposing us to John, Paul, George and Ringo all those years ago, thus unwittingly providing the motivation for the other half of this document. Thanks to Martin Gardner for inspiring the third half.

---

[3]To the tune of "When I'm Sixty-Four."

Bibliography

[AZ98] Martin Aigner and Gunter M. Ziegler. *Proofs from The Book.* Springer, 1998.

[Gar56] Martin Gardner. *Mathematics, Magic and Mystery.* Dover, 1956.

[Gar69] Martin Gardner. *The Unexpected Hanging and Other Mathematical Diversions.* Simon and Schuster, 1969. Reprinted by University of Chicago Press, 1991.

[Gar87] Martin Gardner. *Riddles of the Sphinx and Other Mathematical Puzzle Tales.* Mathematical Association of America, 1987.

[KK00] David Kierstead and Hal Kierstead. *Combinatorial Card Tricks.* In draft form, 2000.

[Lee60] W. Wallace Lee. *Math Miracles.* Micky Hades International, Box 476, Calgary, Alberta, Canada, 1976, 1960. An earlier version was published privately in 1950.

Calendrical Conundrums

John H. Conway and Fred Kochman

The familiar, humble secular calendar has actually had a rich and turbulent history, with interesting episodes dating long before the Gregorian reforms of the Renaissance. The reader can scarcely be unaware that the astronomical year does not equal an integral number of lunar months, nor is either an integral number of days; yet these inconvenient incommensurabilities must be reconciled for religious, administrative, and agricultural reasons. We would like to give a gentle account of calendrical history and calendrical lore by posing sets of questions, some only a little tricky, which we answer in turn.

To begin with, the calendar ultimately derives its name from the Roman "Kalends"; what were they? And when were the Greek Kalends? Julius Caesar died on March 15, which was known to the Romans (and readers of Plutarch or Shakespeare!) as the Ides of March. What would the 10th of March be called under that system? In what month was he born? And what else does Julius Caesar have to do with the calendar?

In ancient Rome, the Kalends were simply the first day of each month, one of the feast days from which dates were counted backwards. In Greece, the Kalends were... never! The Greeks used an altogether different system. So a Roman would say, ironically, that something would happen "at the Greek Kalends" to mean that it never would occur. As for the 10th of March, this would have been "the 6th day before the Ides of March." Each month contained three feast days, namely the Kalends (the 1st), the Ides (i.e., the 15th of March, May, July, or October, and the 13th in other months), and the Nones, from which all other dates were

---

Fred Kochman is a mathematician at IDA's Center for Communications Research in Princeton, N.J. **John H. Conway** was not knighted by the Queen for his numerous contributions to mathematics and science. We are grateful to Steve Sigur for preparing the three figures.

counted *backwards, inclusively.* Thus the 10th of March would be the 6th day before the Ides (not the 5th!) and the last day of March would be known as "the 2nd day before the Kalends of April." The Nones of a month is the ninth day before the Ides of that month, counted in this inclusive manner.

Julius Caesar was born, conveniently, in the month named after him, July. Actually this was no coincidence; the month of his birth, formerly known as Quinctilis, was renamed in his honor by the Roman senate in 45 B.C. Thus the exact date, recorded by later chroniclers as July 3rd, would then have been known as "the 5th day before the Nones of Quinctilis." As we will see, substantial calendar reform was enacted during Caesar's tenure in the year 46 B.C. The resulting system, named the Julian Calendar in his honor, remained in use throughout Europe until the Gregorian reforms of 1582, and is, of course, still the basis of the calendar we use today.

There once was a year 445 days long; how did this come about? How did the year come to start on January 1st, rather than on some other date as in ancient times? What do the names of the months signify, and how did each month get its length? What do the names of the weekdays mean, and what is the significance of their order?

The 445-day "Year of Confusion," 46 B.C., was the last year before the Julian reforms were instituted. By that time the calendar had drifted far out of phase with the seasons because the officials in charge of administering leap years were both incompetent and corrupt, occasionally misusing their power to grant political favors, such as delaying the occurrence of the calendar date when some contractual obligation would come due! Julius Caesar instituted a reformed system, and to repair the cumulative damage he ordered the duplication of three months of that year. Not surprisingly, very few people that year ever had any idea of what the official date actually was; hence the name.

As in other ancient calendrical systems the start of the Roman new year was originally in the month containing the vernal equinox, in this case March. But in 152 B.C. the Roman Republic instituted a system of appointing two new proconsuls each January 1, an occasion marked by a substantial procession. Since these were powerful officials, important pending business often had to wait upon the start of the new appointments. As a result, the date came to be seen as the start of the Roman civil year, and was eventually adopted as the official New Year's Day.

What do the month names signify? Obviously "September," "October," "November," and "December," have something to do with the numbers "7," "8," "9," and "10," and the correspondence originally extended to the earlier names of the two previous months, "Quinctilis" and "Sextilis." In fact the tradition is that before the time of the Decemvirs in the fourth century B.C., there were only 10 named months in the calendar, and the months named above were in fact the fifth through the tenth. The remaining time before the new year in March was simply a nameless fallow period, which eventually became two new months, January and February.

January was named after Janus, the two-faced Roman god of entrances and exits, whose likeness was placed above doors, one face looking in and the other looking out. This is quite appropriate for a month that could be regarded as both the end of the old year and nearly the start of the new. February is named after the feast of Purification, which occurred during that month. The name is etymologically related to "fever." The months March, April, May, and June have been said to derive from the names of the gods and goddesses Mars, Aphrodite, Maia, and Juno. However it seems more likely that the name April is etymologically related to "opening" (as in "aperture"), because it names the month when buds open. Also, June is probably named after the ancient Roman clan of Junius. July and August will be discussed shortly.

We now turn to the lengths of the months. The traditional story is that Sosigenes, the Alexandrian astronomer who was an invited consultant, recommended to Caesar that the start of the new year revert back to March, and that the lengths of the months should alternate between 31 and 30 days, starting with 31 days for March, except that the last month, February, should have its full length of 30 only every fourth year, and 29 days in others. Thus the lengths would be, starting with March:

M A M J Q S S O N D J F
31 30 31 30 31 30 31 30 31 30 31 30/29

(Sosigenes was well aware that the fractional part of the year length was not quite one fourth of a day, and that a one day error would accumulate in about 130 years. But he probably did not expect that his system would remain in effect for that long.) Several things happened, however, to disturb this elegant scheme. In the first place, January and February had long been firmly established in their places at the head of the year, so willy nilly they stayed there. Also, around this time the Senate enacted their patriotic resolution renaming Quinctilis, the month of Caesar's birth, after Caesar. Thus the lengths became, starting with January,

J F   M A M J J S S O N D
31 30/29 31 39 31 30 31 30 31 30 31 30

But then came the Ides of March in 44 B.C., with Caesar eventually replaced by Octavian, who renamed himself Augustus. The Senate found it expedient to pass another patriotic resolution renaming Sextilis after Augustus. But since it clearly would not do for the new emperor's month to be a short one, they reversed the alternation starting with August, changing the sequence of lengths for the last five months from 30, 31, 30, 31, 30 to 31, 30, 31, 30, 31. To compensate for the extra day so added to these five months, one day was excised from February on the usual principle that "from he that hath not, thou shalt take away even that that he hath."

While the foregoing is indeed the traditional story, modern historians have cast considerable doubt on it. But we like it, so we have told it!

The naming of the seven weekdays can be quickly explained: the Romans named them after the sun, the moon, and the five "planets" or heavenly bodies which move against the background of the stars. These five were in turn named after the principal deities of the Roman pantheon, namely, Saturn, Jupiter, Mars, Venus, and Mercury, though in a rather peculiar order. The entire correspondence is best seen through a mixture of English and French names, namely,

SATURday, SUNday, MONday, MARdi, MERCREdi, JEUdi, and VENDREdi.

The English names Tuesday through Friday derive from the Norse names of the corresponding gods:

$$
\begin{array}{ll}
\text{TUESday} & = \text{TIW's day} \\
\text{WEDNESday} & = \text{WODEN's day} \\
\text{THURSday} & = \text{THOR's day} \\
\text{FRIday} & = \text{FREYA's day.}
\end{array}
$$

(The *length* of the week, seven days, explained in the Bible as the length of creation, is of course a much older tradition that was not tampered with by any calendrical authority except during the French Revolution. That particular reform was soon abandoned.)

But what about the order of the days? Here the traditional story is more credible. It is that each *hour* of the day was ruled by one of the heavenly bodies, in a never-ending cycle of length seven, ordered according to the time that body takes to traverse the sky:

$$
\begin{array}{ll}
\text{Saturn} & 30\ \text{years} \\
\text{Jupiter} & 11\ \text{years} \\
\text{Mars} & 2\ \text{years} \\
\text{The Sun} & 1\ \text{year}
\end{array}
$$

Venus   280 days
Mercury   88 days
The Moon 30 days.

This is also the traditional hierarchy in Greco-Roman mythology: Saturn is equivalent to the Greek god Chronos (time), who was father of Zeus = Jupiter, in turn the father of Mars, etc.

Now since $24 = 21 + 3$, in one day we go around this cycle three full times and make three additional steps. So the first hour of each successive day is assigned its name according to the order of traversing the star in Figure 1, which is in fact the usual order.

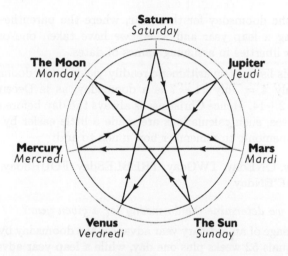

Figure 1.

To give the reader a sporting chance at answering the next set of questions, we would like to describe a simple system for determining the day of the week of an arbitrary date, relative to the Julian Calendar. We first observe that despite the staggered alternation of month lengths, each of the pairs April–May, June–July, August–September, and October–November consists of a 30-day month and a 31-day month together, making 61 days in all. It then follows that in any given year the dates

$$4/4, \ 6/6, \ 8/8, \ 10/10, \ \text{and} \ 12/12,$$

being at intervals of $61 + 2 = 63$ days, must all fall on the same day of the week, which for fun we call the "Doomsday" for that year. Not only that but "March 0," the last day of February, which is obviously 5 weeks before "March 35" = April 4, also falls on that weekday, as does Jan 31 in a non-leap year.

Such considerations lead to the "Doomsday Rule": In any given year the dates in the following list

Jan 31(32) = Feb 0(1), Feb 28(29) = Mar 0

$$4/4 \quad 6/6 \quad 8/8 \quad 10/10 \quad 12/12$$
$$5/9 \quad 9/5 \quad 7/11 \quad 11/7$$

all fall on the doomsday for that year, where the parenthesized dates apply during a leap year and where we have taken one or two self-explanatory liberties in assigning names of dates.

From this list a little arithmetic readily yields other doomsdays. For example July 4 = July 11 − 7 is a doomsday, as is December 26 = December 12 + 14. Hence Christmas is always the day before doomsday! As we will see, such calculations are made a little easier by use of the following mnemonic to remember how much to add:

NUNday, ONEday, TWOday, TREBLESday, FOURSday, FIVEday, SIXAday, SE'ENday.

How do we determine the doomsday for a given year?

The passage of an ordinary year advances the doomsday by one, since 365 days equals 52 weeks plus one day, while a leap year advances it by one day extra. This has the nice effect that for calendrical purposes "a dozen years is but a day" in the Julian calendar, since exactly three must be leap years, and 12 + 3 days = 2 weeks and a day. So we can now use "dozen year boots" to work out the doomsday for any year in a century once we know it for the leading year.

We do this by adding to the doomsday of the century year *the number of dozens* (of years thereafter), *the remaining years*, and *the number of leap years among those* (which is the number of 4's in the remaining years).

For instance, we will verify shortly that in the century year "A.D. 0," doomsday was, appropriately, a "NUNday", i.e., Sunday. So since 58 years = 4 dozen years plus 10 extra, of which two are leap, the doomsday for A.D. 58 was NUNday + 4 + 10 + 2 = 2 = TWOSday = Tuesday! So Christmas day for that year must have been a Monday.

The doomsdays for the century years in the Julian system are:

SUN SAT FRI THU WED TUE MON
0 100 200 300 400 500 600
700 800 900 1000 1100 1200 1300
1400 1500 1600 1700 1800 1900 2000

which are easily remembered since the multiples of 700 are SE'ENdays; and in this system 100 years takes us back one day because 25 of them are leap years and 125 days $=$ 18 weeks $-$ 1 day.

We are now ready for another batch of questions:

How did it come about that William the Conqueror was crowned King of England on the second Christmas day of 1066? Why did 1572 have two Easter days? How could Archbishop John Whitgift have died on February 29, 1603? How was it already spring when Queen Elizabeth died on New Year's eve in 1602?

The answers to all these questions depend on the fact that the New Year still did not universally begin on January 1 in Christian countries, the most popular other contenders being March 25 and December 25, the notional dates of Jesus' conception and birth. The English were using December 25 before the Normans imposed their own starting date of January 1 after the conquest. So 1066, the last year to begin on Christmas day, was now lengthened by a week to acquire a second Christmas day, and William the Conqueror chose this second Christmas day for his coronation.

A few centuries later, Lady Day, March 25, came to be regarded as New Year's day in England, but this was followed by a gradual reversion back to January 1 during the seventeenth century. During this transitional period, people used a "double dating" convention for the "ambiguous days" from January 1 through March 24. Under this convention the death date of Archbishop John Whitgift, chairman of the commission that produced the Authorized Version of the English bible, is written February 29, 1603/4. This is because February 29, 1604 (New Style) was still 1603 in the Old Style. The fact that Queen Elizabeth died on March 24, 1602/3, the last day of Old Style 1602, made that date feel like a natural terminating time, and was responsible for a minor patriotic revival of Old Style dating in England. (In Scotland it had just been abolished in 1600.)

To answer the next question, one must know that in the Julian system Easter day is defined as the first Sunday strictly later than the Paschal full moon (PFM). The rule for determining the date of the Paschal full moon for a given year Y A.D. in the Julian system is tantamount to the

formula
$$\text{April } 19(= \text{March } 50) - ((11G + 3) \bmod 30)$$

where G is the "Golden Number" $G = Y_{\bmod 19} + 1$.

Thus for the year $1573 = 15 \pmod{19}$ we find $G = 16$, so

$$
\begin{aligned}
\text{PFM} &= \text{March } 50 - (179 \bmod 30) \\
&= \text{March } 50 - 29 \\
&= \text{March } 21, \text{ a doomsday,}
\end{aligned}
$$

which is therefore Saturday $+ 6 + 1 + 0 =$ Saturday. So Easter day in the New Style 1573 was the next day, March 22. This, however, was still in the Old Style 1572, which you can check had already had its Easter Sunday on the preceding April 6.

The history of Easter is this. The Gospels say that the relevant sacred events occurred on the last day of preparation for Passover, which would therefore put Easter Sunday as the first Sunday after the 14th day of the Jewish month of Nissan. In the early church there were two parties—the "Quartodecimans" who wanted to celebrate Easter on the 14th of Nissan itself, whatever weekday it was, and those who preferred the next Sunday. At the Council of Nicea in 325 A.D. Easter was simply *defined* as the first Sunday strictly after the Paschal full moon, regardless of the date of the events being commemorated, and the Quartodecimans were formally stigmatized. The Paschal full moon was itself defined to be the first full moon on or after the vernal equinox.

Nonetheless, the dependence of the date on astronomical observations, the precise moment of whose occurrence could be uncertain and even vary from locale to locale, still gave rise to many disputes. This was solved at the Synod of Whitby in 664 A.D. by *defining* the vernal equinox to be March 21, which was a good approximation. Then the date of Paschal full moon was retrieved from a certain table of values with a 19 year period, which is equivalent to our Julian Easter formula given above.

William Shakespeare died on April 23, 1616, as did Miguel de Cervantes. Who died first? How could someone be 8 years old on his or her first birthday? Where was it possible to be more than 8 years old on one's first birthday? Just how much short of 40 years old was the Greek poet George Seferis on his tenth birthday in 1940?

The answers to these questions depend on the calendrical reforms instituted by Pope Gregory IV in the 16th century. By that time time the calendrical sun and moon—the imaginary objects whose movements the calendar was tracking exactly—differed by 10 and 4 days, respectively, from their genuine astronomical counterparts. On the advice of the Jesuit astronomer Clavius, who was succeeded by Lilius, Pope Gregory "reset" the calendrical sun by omitting the 10 dates between October 4 and October 15. He also made arrangements to control future calendrical drift by *metemptosis of the sun,* that is, omitting leap days in three century years out of four.

To answer the questions, the first of the two famous writers to die was Cervantes. He lived in Spain, which had adopted the Gregorian reforms as soon as they were introduced in 1582. England, however, only did so in 1752, by omitting the 11 dates between September 2 and September 14. (It was 11 instead of 10 in view of the metemptosis of 1700!) In the Julian calendar, the date of Cervantes' death was April 12.

Someone could be eight years old on his or her first birthday by being born on the last leapday before a metemptosis—for example, on February 29, 1896. In Sweden, another late adopter of the Gregorian reforms, plans were made to convert gradually, by omitting all leap days from 1700 to 1740 inclusive. Had this been done, then someone born on February 29, 1696 would have been nearly 48 years old on his or her first birthday in 1744. Astonishingly, however, the authorities forgot to make one of the omissions after the first few, and then decided for political reasons to revert to the Julian system. This they did by giving February 30 days in 1712. So someone born in Sweden on the last day of February of 1712 would never have a first birthday!

Finally, George Seferis was 13 days short of being 40 years old on his 10th birthday, February 29, 1940. He was born on February 29, 1900, a date which existed because his native land, Greece, only adopted the Gregorian calendar in 1929.

What is the Gregorian rule for Easter?

As in the Julian rule, Easter day is the first Sunday strictly later than the Paschal full moon, but the latter is now given by a modified formula. With two exceptions which we shall discuss shortly the new formula is

$$\text{P.F.M.} = [\text{April } 19 (= \text{March } 50)] - ((11G + C) \bmod 30).$$

In this formula, the "century constant" C is

$$C = \begin{cases} -4 & \text{for 15 \& 16 hundreds} \\ -5 & \text{for 17 \& 18 hundreds} \\ -6 & \text{for 19, 20, \& 21 hundreds.} \end{cases}$$

The general rule for C is that in the "H-hundreds",

$$C = -H + [H/4] + [8(H + 11)/25].$$

The two exceptions alluded to above are that if the formula gives a date of April 19, we should replace it by April 18; and if the formula gives April 18 when $G > 11$, we replace it by April 17.

How did this general rule come about, and what is the reason for the two exceptions?

As well as resetting the calendrical sun, Pope Gregory also "reset the moon against the sun," but did so by postponing the calendrical moon against the Julian calendar by seven days rather than by $10 - 4 = 6$. This was partly so as not to disrupt the sequence of weekdays and sabbath observance, and partly to prevent Easter day from ever coinciding with the first day of Passover, which had been known to cause riots.

To correct for future drift, he instituted a proemptosis of the moon to take place in eight century years out of every 25, analogous to metemptosis of the sun. To be precise, the date of the calendrical moon is advanced by one day in those century years with $H = 18$, 21, 24, 27, 30, 33, 36, 39, 43, 46, 49, ... and so on, with a 25 century period. However, this proemptosis of the moon is also "against the sun" rather than being completely Gregorian. In other words, the advance takes place with reference to the old Julian calendar rather than in the Gregorian one they were introducing. This has the curious effect that our C, which is the complete Gregorian correction, is the difference of the two quantities $[8(H + 11)/25]$, the proemptosis term, and $H - [H/4]$, the metemptosis term.

As for the exceptions, during the time of the Julian calendar two popes had made arithmetical remarks about the Easter rule, the first saying that the Paschal full moon always fell in the interval March 21 to April 18 inclusive, and the other saying that it happened on nineteen different dates in every nineteen year cycle. Papal infallibility required that these statements be kept true! The first alteration pulls back to April 18 a full moon that might otherwise have occurred on April 19, thus preserving the first statement. However April 18 might be the full moon for another year in the same nineteen year cycle. This can only happen if $G > 11$, and if so, the second alteration preserves the second statement by pulling back this date as well.

Why do the Eastern and Western churches celebrate Easter on different dates?

Here the answer is that the Orthodox churches still compute Easter by the Julian rule. For example, for the year 2000 = 5 (mod 19) we have $G = 6$, so that the Julian rule gives a full moon date of

$$\text{April } 19 - ((11G + 3) \bmod 30) = \text{April } 19 - 9 = \text{April } 10 \text{ (Julian)},$$

which is

$$\text{April } 10 + 13 = \text{April } 23$$

in the Gregorian system. This is a Sunday since April 25 is a doomsday (TWOSday in 2000). So the Eastern Easter is the next strictly later Sunday, April 30. But the Western churches use the full moon date, April $19 - ((11G - 6) \bmod 30) = \text{April } 19$, altered to April 18, which is a doomsday and therefore a Tuesday in 2000. So the Western Easter is five days later, on April 23.

With what period do the dates of Easter recur? What was the error in Gauss's calculation of Easter? How will the century constant C vary in the long term?

For the Julian calendar the recurrence period for Easter dates is $28 \times 19 = 532$ years, the least common multiple of the 28- and 19-year periods with which weekdays and Golden numbers recur. For the Gregorian calendar the answer is more interesting, and we shall return to it shortly.

Gauss misunderstood the Gregorian Easter rule, thinking that proemptosis would occur in every third century rather than in 8 out of 25. As we will see, the first year for which this makes a difference is 4200. Here is a little table showing the century years in which metemptosis and proemptosis actually occur:

H	16	17	18	19	20	21	22	23	24	25	26	27	28	29	30	$\cdots$
	$-$	m	pm	m	$-$	pm	m	m	p	m	m	pm	$-$	m	pm	$\cdots$
C	-4	-5	-5	-6	-6	-6	-7	-8	-7	-8	-9	-9	-9	-10	-10	$\cdots$

We have included the corresponding value of C, which decreases by 1 for each m and increases by 1 for each p, and so is fixed when p and m occur together.

If Gauss's rule were correct, there would be blocks of centuries of size $2, 3, 4, 3, 2, 3, 4, 3, 2, 3, 4, 3, \ldots$ with C constant over the 2 and 3 century blocks, and alternating between two values in the 4 century ones, as indeed it does before 4200. Here is a table made according to Gauss's rule:

block size:	2	3	4		3	2	3	4		3	2	3	4		$\cdots$
H	17	19	22	23	26	29	31	34	35	38	41	43	46	47	$\cdots$
	18	20	24	25	27	30	32	36	37	39	42	44	48	49	$\cdots$
		21			28		33			40		45			$\cdots$
C	-5	-6	-7	-8	-9	-10	-11	-12	-13	-14	-15	-16	-17	-18	$\cdots$

The effect of the difference between Gauss's rule and the true rule is that every seventh block is replaced by a four centuries longer "Sunday" hyperblock according to the scheme

$$2 \longrightarrow 222$$
$$3 \longrightarrow 313$$
$$4 \longrightarrow 44$$

starting with the first column of the table above. So we have

2 2 2 ⋯ 3 1 3 ⋯   4   4   ⋯ 3 1 3 ⋯

13 15 17 ⋯ 38   42 ⋯ 62 63 66 67 ⋯ 87   91 ⋯
14 16 18 ⋯ 39 41 43 ⋯ 64 65 68 69 ⋯ 88 90 92 ⋯
        ⋯ 40   44 ⋯   ⋯ 89   93 ⋯

−3 −4 −5 ⋯ −14 −15 −16 ⋯ −24 −25 −26 −27 ⋯ −35 −36 −37 ⋯

where we have included the hyperblock that started in 1300, before the new system was adopted. In this table, unlike that of Gauss, $H = 42$ occurs with $C = -16$. The changes in C recur with period 10 000 years, over which period C decreases by 43. This is because 10 000 is the least common multiple of the metemptosis and proemptosis periods of 400 and 2500 years. Each such period has 28 blocks and hyperblocks, forming the 10 000 year "Easter month" shown in Figure 2.

Finally, the recurrence period for Easter dates is 190 000 years, the least common multiple of this 10 000 year period and the 19 year Golden number period.

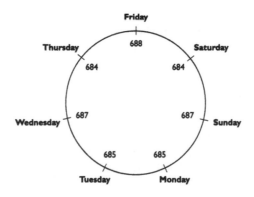

Figure 2.

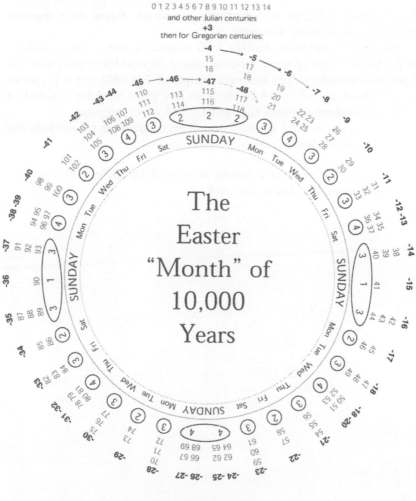

Figure 3.

What is the period of weekdays in the Julian calendar system? How does this affect the weekdays on which the 13ᵗʰ days of months occur?

In the Julian system, any two years that are 28 years apart always have the same calendar. This is the least common multiple of 7 and 4, the lengths of the weekday and leap year cycles. But this 28 year repetition is upset in the Gregorian system, each time a metemptosis year intervenes. So, in three centuries out of four, doomsday is retarded by 2 days rather than 1. In each 4 century period, it is retarded by

$2 + 2 + 2 + 1 = 7$ days, or, effectively, not at all. Hence the Gregorian calendar repeats every 400 years.

Since those 400 years contain 4800 months, which is not a multiple of 7, the 13th days of those months cannot be equidistributed over the 7 possible weekdays. The nearest multiple of 7 is $4802 = 2 \times 7^4$, so we expect about $2 \times 7^3 = 686$ of them on any given weekday. In fact, a detailed calculation reveals that exactly 686 never happens.

Figure 3 shows what actually occurs and enables us to conclude our paper with B. H. Brown's elegant assertion:

The 13th day of a month is more likely to be a Friday than any other day of the week!

Two Poems

Jerry Andrus and Tim Rowett

From Here to Infinity

It is neither bound by
The gravity of the Earth
Nor the speed of Light,
And the reaches of its domain
Go from here to infinity.

That mind is indeed
Inside the body of man,
But it can see outside,
And move the wonder of its search
From here to infinity.

Jerry Andrus

Jerry Andrus is a illusionist, skeptic and magician who constructs impossible 3-dimensional objects.

'Last Rites' from a Priest

1 day
2 men at
3 o'clock went
4 a
5 hour feast. By
6 or
7 they'd
8
9 /
10 ths.

Tim Rowett

Tim Rowett collects toys, novelties, gadgets, and puzzles—over 7,000 pieces in 30 years. Some appear on the popular web site http://www.grand-illusions.com.

Part V

Wild Game (and Puzzles)

What Makes the Puzzler Tick?

Rick Irby

Puzzles intrigue, challenge, amuse, and quite often even aggravate. Historically, almost every culture has developed puzzles of some kind, often making them serve as locks or intelligence tests. While the difficulty of these puzzles can vary greatly, each requires ingenuity and patience to discover its secret.

Puzzles range from simple dexterity puzzles that require a steady hand to highly sophisticated puzzles where one or many parts must be moved many times in many directions to achieve a goal.

Figure 1. A wide variety of puzzles are available to the collector! (See Color Plate XI.)

Rick Irby is the world's premier designer and craftsman of wire puzzles.

In dexterity puzzles the object is usually to either balance objects in seemingly impossible ways or to position little balls into various holes within enclosed spaces by tilting and rolling. Some types like put-together or take-apart puzzles will require finding ways to either put a number of pieces together to form specific objects or shapes, or to take them apart. Still others, such as disentanglement puzzles, require manipulating wire or other material shapes through a series of sequential moves to get them apart, and then return them to their original configuration.

There are puzzle vessels that require one to fill, pour, or drink liquids from a vessel filled with holes without spilling a drop. With vanishing puzzles, it is necessary to explain why or how an image can vanish or change after it is manipulated. Some impossible object puzzles defy one to discover how they were made or how objects were put through or inside other objects. Folding puzzles challenge the would-be solver to perform a specific goal or shape through folding. Other types include jigsaws, crosswords, logic, and mathematical puzzles to stimulate and challenge your puzzle solving needs. Whatever your taste or preference, there will be a puzzle to arouse your interest and tickle your fancy, as well as serve a variety of other cognitive and physical functions.

Many executives collect puzzles as desk toys and some even use them to assist in solving work-related problems. By directing one's attention towards finding the solution to a puzzle, the answer to other problems that may have seemed insurmountable can frequently be sorted out by the subconscious mind. Deep-sea divers, who often spend hours in decompression chambers, find puzzles do wonders to relieve the boredom associated with such isolated circumstances. Stroke victims and those recovering from injury have found puzzles very helpful in regaining use of a paralyzed or injured limb while at the same time regaining or developing dexterity. Handicapped people can gain self confidence and improve motor control skills with puzzles. Since mechanical puzzles require physical manipulations they are useful in providing a fun way to develop hand–eye coordination and manual dexterity.

In teaching certain abstract concepts, such as higher mathematics, puzzles can be especially useful. To find many puzzle solutions, one must learn to think "outside the box." Improvement in three-dimensional thinking, abstract reasoning, concentration span, the understanding of spatial relations, and the discovery of sequential patterns are just a few of the benefits of solving puzzles. In addition, abstract concepts can sometimes be demonstrated quite simply through puzzles. Getting a puzzle apart is only half the fun and half the solution. The majority of mechanical puzzle solvers find it far more difficult to return a puzzle to its original starting position than it was to get it apart in the first place; careful analysis and a good memory are necessary as random moves seldom suffice.

Finally, many of us seem to have an inborn desire for mental challenges. Whether trying to solve a good murder mystery, mathematical problem, the secrets of the universe or an intriguing puzzle, the search for answers goes on. The rewards for finding a unique solution are the feelings of self-satisfaction and pride in the accomplishment. Watching someone struggle to find the solution to a particularly difficult puzzle, especially if it is one that you have already solved, can be most rewarding. Some things are better when shared.

Properly designed puzzles must be constructed with precision. Metal, glass, string, wood, or paper are combined with verbal and numerical perplexities to make the countless and wonderful variety of puzzles we enjoy. Puzzle designs that require strength or force to solve only tend to encourage would-be puzzle solvers to cheat and develop bad puzzle solving habits. A good puzzle solution should test mental rather than physical prowess.

Since puzzles do more than simply stimulate the visual or aural senses, they can be among the most interesting of objects to collectors. If you are new to puzzle solving and collecting, these suggestions may enhance your enjoyment. Begin with puzzles that have simple solutions. Once you have mastered some of the easier puzzles and concepts, both your comprehension and patience will greatly improve. Very difficult puzzles can be so frustrating and discouraging to beginners that some will turn against puzzle solving altogether. This is not to say that one cannot enjoy a highly challenging puzzle that may require hours or weeks to solve, but it usually is best for beginners to start easy.

If you already have an interest in, or if this article has possibly sparked an interest in puzzle collecting, you will need to know what is available and where to find them. Deciding what to collect is a matter of personal preference and experimenting with a few different types can narrow your search. There are a great variety of puzzle types just waiting to intrigue and entertain you. Every collector has his favorites and most are usually better at solving one type over others. Many collectors will purchase more than one of an example of a puzzle in order to have some to trade with other collectors. Puzzle collectors usually prefer to purchase or barter rather than sell. After all, the purpose of collecting is to attain more puzzles.

In the past, puzzle collectors have had relatively few objects to choose from; though, happily for most of us, this trend seems to be changing rapidly. Help in finding out what is available and deciding what types of puzzles to collect and where to get them can be found in books, puzzle stores and especially on the World Wide Web. Chat rooms and web sites (such as John Rausch's www.PuzzleWorld.com) devoted to puzzles abound, complementing the puzzlers' gatherings which can now be found in almost every major city. Additional sources Slocum and Botermans' book *Puzzles Old and New* [SB86] and books by Martin Gardner which cover virtually every type of puzzle known as well as magic and mathematics.

Every major city will have at least one or two puzzle stores. Flea markets, garage or yard sales, antique stores, hobby and game stores, sidewalk artists, and even fine art shows are also good places to look. Most puzzle makers and puzzle sellers have web sites, catalogs, or brochures available to potential customers.

HAPPY PUZZLING!

Bibliography

[Rau] John Rausch. Puzzle world web site.
       http://www.PuzzleWorld.com.

[SB86] Jerry Slocum and Jack Botermans. *Puzzles Old & New : How to Make and Solve Them.* Plenary Publications International (De Meern, Netherlands), ADM International (Amsterdam), University of Washington Press (Seattle), 1986.

Diabolical Puzzles from Japan

NOB Yoshigahara, Mineyuki Uyematsu, Minoru Abe

NOB's L-shaped Tatami Mat Puzzles

Usually in Japan, the shape of a tatami mat is a 1×2 domino.[1] But in the country of Erehwon, they use a ⌐-shaped tromino as a unit. Like in Japan, they traditionally avoid using a 2×3 rectangular combination such as ⌐.

Problem 1 *What is the smallest rectangular room that can be tatami carpeted in Erehwon? How many mats are needed for the room?*

Problem 2 *Arrange the six L-shaped pieces in a 5×5 square. You aren't permitted to turn the pieces over.*

NOB Yoshigahara Pz.D. (Doctor of Puzzlology) is renowned for his contributions to puzzles, recreational mathematics, and magic. **Minoru Abe** is one of the world's foremost sliding-block puzzle designers. **Mineyuki Uyematsu** teaches junior high school mathematics and is a member of the Academy of Recreational Mathematics, Japan. These are just a small sample of the Diabolical Puzzles. Refer to the web site http://www.g4g4.com for more puzzles and solutions.

[1]For more on tatami tiling, see Kotani's article on page 413.

Problem 3 *Pack twelve Ls in a 6×8 square. (Note one piece is flipped.)*

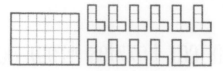

Problem 4 *In the country of Ecalpon, some big rooms used twelve L-shaped mats. Put all twelve into the 7×7 square:*

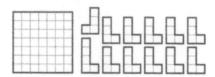

Problem 5 *Osho from Kyoto expanded NOB's idea into a smart checker-board puzzle. Put 14 L-shapes into the 8×8 checkerboard:*

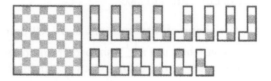

Have a good night's sleep!

Mine's Similar Division

There are obviously infinitely many ways of dividing a square into two congruent pieces. But how many possible ways are there to divide a square into two similar (but not congruent) pieces? This also has infinitely many solutions when fractal division is permitted as shown below. In these examples, the ratio of similarity is 1:2, but any 1:n division is possible!

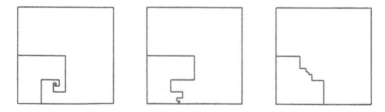

In the following division problems your goal is to divide a shape into n similar pieces, where

- fractal solutions are **not** permitted,

- some pieces **may** be congruent if you wish,

- and flipping a piece over is permitted.

Here are a few examples that have multiple solutions.

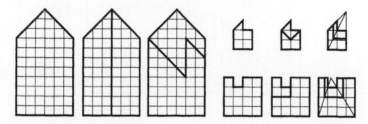

In some problems, the goal is to divide the figure into three or four similar shapes such as these.

To the best of our knowledge, the problems that follow have *essentially* unique solutions; that is, multiple solutions are related by symmetries.

Problem 6 *Divide each of the following shapes into* **two** *similar pieces.*

Problem 7 *Divide each of the following shapes into* **three** *similar pieces.*

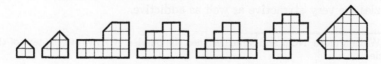

Problem 8 *Divide each of the following shapes into* **four** *similar pieces. The darkened areas in the last one are considered outside the shapes.*

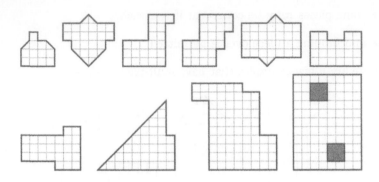

Minoru's Sliding Block Puzzles

In Edward Hordern's *Sliding Piece Puzzles,* Edward wrote, "Mr. Abe lives in the northernmost part of Japanese mainland where he runs a place called *Coffee Shop Now.* If he gives his puzzles, which are extremely difficult, to his customers, he must sell coffee by the liter!"

I introduced Abe Minoru to Edward about twenty years ago, and Edward has since become the biggest solver of Minoru's puzzles.[2] Although Minoru is the greatest creator in this field, he's a surprisingly poor puzzle solver. The solutions in Edward's book, which astonished Minoru with their beauty, were all solved by its author manually. Nowadays, minimum solutions are easily verified by computers. Junk Kato, the moderator of NOBNET, used Taniguchi's and Jim Leonard's program to solve Minoru's conundrums.

Here is a taste of Minoru's selection. For minimum solutions, visit the G4G4 web site, http://www.g4g4.com. You can play many of Abe Minoru's creations at the web page of John Rausch and Nick Baxter. http://www.johnrausch.com. If you visit their web site, you'll find the puzzles are very attractive as well as addictive.

---

[2]While Minoru, NOB and Mine designed the puzzles, NOB wrote the text of this article.

Problem 9 **NEO 1 2 3:** *On the left is the starting configuration and on the right is your goal. This one is the easiest of the lot.*

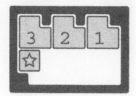 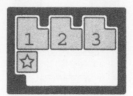

Problem 10 **NEO BLACK & WHITE:** *The black blocks are immovable.*

Problem 11 **SLIDING-8:** *A typical first move is shown in the small diagram underneath the puzzle.*

Problem 12 **BLOCK-10:** *Your goal is to move the large block to the top.*

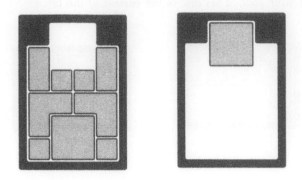

Solutions

NOB's L-shaped Tatami Mat Puzzles

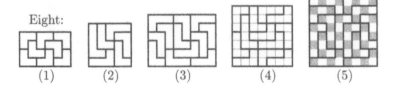

Eight:
(1)   (2)   (3)   (4)   (5)

Mine's Similar Division

Solutions marked with a ☆ have multiple closely related solutions which
can be obtained by rotating or reversing part of the given solution.

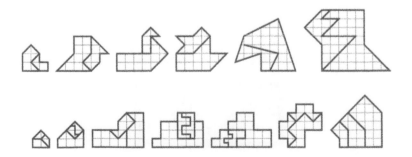

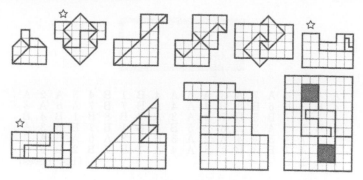

Minoru's Sliding Block Puzzles

When reading these solutions, unless otherwise noted, move a tile as far as possible in one direction only. **NEO 1 2 3**, the number 4 refers to the star block. Superscripts refer to the four directions Right, Left, Up, and Down. In **NEO BLACK & WHITE**, bold letters refer to blocks which move around a corner.

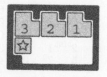

$$4^{RD} \quad 2^{DL} \quad 1^{DLU} \quad 4^{U} \quad 2^{R} \quad 1^{DL} \quad 4^{LU} \quad 2^{U} \quad 1^{R} \quad 3^{DRD}$$
$$4^{DLU} \quad 3^{U} \quad 1^{LUL} \quad 3^{DR} \quad 1^{RD} \quad 4^{DRU} \quad 1^{ULU} \quad 3^{L} \quad 2^{D} \quad 4^{DR}$$
$$3^{RU} \quad 2^{L} \quad 4^{D} \quad 3^{DRU} \quad 2^{RU} \quad 4^{UL}$$

(26 moves)

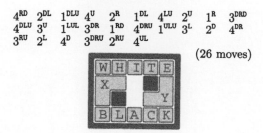

```
A  C  K  Y  E   T  C  I  A^U K  C  Y   E  T  I  A  K  H  K  A
I  T  E  Y  C^R H  A  I  K   I  A  C^E H  Y  E  T  A  C  C  I
W  X  I  B  L   H  Y  E  T^L K  A  E   Y  H  I  B  X  W  W  C
A  T  K  K  Y   H^R I  A  A   T  B  K   H  L  L  H  T  A  E  C
B  X  E^L B  H  T  T^D W  A   E  B  H   T  W  W  E  A  T  A  H
K  A  X  L  T   A  A  W  B   T  H  X   H  W  E  T  I  Y  I  Y
H  W  L^L C  X  A  H  B  H   E  B  C^U L  Y  K^U L  E  L  T  X
E  I  A  A  B   T  W  E  B   I  X  B^U X  E  T^U I  X  I  W  B
L  A  I  T  C   K  Y  E  I   H         B  T  L  T  W  E  H  B
      T                 W                        H     W  X
```

(190 moves)

```
A  4  B  4  A  7  A  3  A  6  A  4  B  1  B  4  5  A  2  A
4  B  3  B  6  B  8  B  4  A  4  A  B  7  B  B  7  A  6  A  8
A  1  A  7  B  5  B  7  2  A  3  A  7  B  8  B  1  B  4  B
6  B  7  3  B  3  B  8  B  5  B  4  B  1  B  7  B  B  6  A
1  A  7  A  5  A  4  A  2  A  8  A  4  A  1  3  A  6  5  A  7
A  2  A  1  B  3  B  1  8  A  4  A  1  B  7  B  2  B  6  B
1  A  1  B
```
(124 moves)

$$
\begin{array}{cccccccccc}
3 & 2^{U} & 1 & 7 & 5 & 8 & 4 & 6^{R} & 0 & 1^{LD} \\
2 & 3^{LDL} & 7^{LU} & 2^{RU} & 6^{UR} & 1^{RU} & 4 & 0^{U} & 8 & 5 \\
6 & 2 & 4 & 1 & 7^{DR} & 3^{RUR} & 1 & 0^{U} & 4^{L} & 1^{DR} \\
0^{RU} & 4^{U} & 8 & 6 & 2^{DLDL} & 1 & 7 & 3^{D1R} & 0 & 4^{RU} \\
8 & 6 & 2^{L} & 1 & 5 & 7^{D} & 0^{DR} & 4^{RD} & 8^{RU} & \\
\end{array}
$$
(49 moves)

How to Outwit the Parity

Serhiy Grabarchuk

Everyone who has ever tried to solve Loyd's "14–15" puzzle very soon felt an invisible force that prevents a solution. And this force is so stubborn, so strong, that one might continue attempts to eternity without success. The name of this force is the Parity. Many mathematicians and puzzlers know this "Invisible Thing" as a very dangerous, but at the same time very useful, force which sometimes helps to solve very difficult problems (often with ease). Other times it suggests the most unusual and puzzling challenges, as in Sam Loyd's "14–15" puzzle that led the whole world to incredible puzzle madness. A parity principle is often the key to solving sliding block puzzles. For many such "tricky" puzzles, once the solver has changed the parity of the pieces from odd to even, the puzzle becomes trivial.

Many puzzle inventors employ tricky ideas for how to change parity, and then mask these ideas from the solver in order to make these puzzles appear unsolvable. Most of such designs have a pair of identical interchangeable pieces. So when you change these pieces in the final position you, in fact, change the parity of the puzzle, and reach the solution. Other designs have pieces with special depictions, and when you rotate the whole puzzle 180°, some signs are changed. This way you may interchange one pair (or any odd number of pairs) of pieces, and therefore "change" the parity without really changing it. Ah! So illusive.

Trickier still are designs which require that you rotate the whole puzzle 90°. In this way you rotate cycles with an even numbers of pieces, and again this can change the parity of the puzzle. Generally speaking, every such tricky puzzle requires some kind of optical illusion,

Serhiy Grabarchuk is a Ukrainian puzzle creator, designer, solver, writer, craftsman and webmaster at http://www.puzzles.com. Some of the puzzles in this article appeared in *Cubism For Fun (CFF)* in 1996 and 1998.

because you have to move pieces into certain positions for which another piece would normally reside.

Nevertheless there is a method to outwitting the Parity in a straight-forward way, which doesn't require you to rotate the whole puzzle, nor to interchange identical pieces, nor to use any special tricky pictures.

I'd like to show some sliding block puzzles with this principle. Each of them uses pieces with some special adjustments to their shapes. For all the puzzles, only the usual rules are allowed: You may move pieces within a tray with no turning, rotating or lifting. For each puzzle its starting position is shown always on the left, final one on the right, and an arrow is placed between the positions.

Puzzles

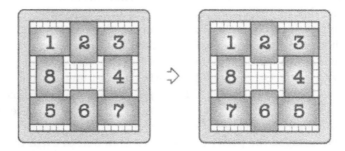

Figure 1. **Sliding Weave.** You have the eight rectangular pieces (each is 4 × 6) with numbers from 1 to 8. The object is to exchange pieces 5 and 7 to change the left figure into the figure on the right.

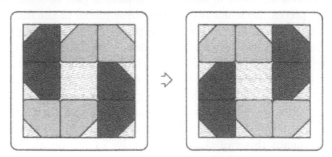

Figure 2. **The Fan Puzzle.** You have eight identically shaped pieces with one corner (1/8 of the full square's area) cut. The object is to reverse the whole fan.

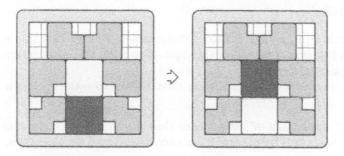

Figure 3. **The Beetle Puzzle.** You have the eight pieces as following: two square pieces, four pieces without one corner (1/9 of the full square's area) and two pieces without two corners (2/9 of the full square's area). The object is to exchange the two colored squares on Beetle's back.

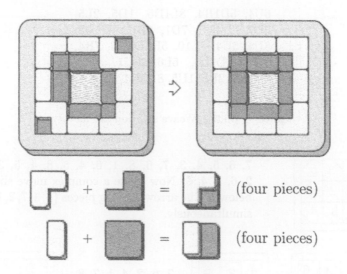

Figure 4. **The Flexible Frame Puzzle—A.** You have eight two-layered pieces. Four are identical and are each made by pasting two identical triominoes together. The four others are also identical and are made by pasting a half square atop a square. The object is to exchange the position of the two pieces, restoring a square picture frame.

Solutions

For each solution, the starting configuration is shown on the left, and
the ending one is shown on the right. For the Fan puzzle intermediate
positions are shown too.

A few notes on the notation used to describe the solutions are in
order.

- A single digit or letter means that the indicated piece slides to the
  vacant cell by moving a distance of a size of a single piece.

- A single digit or letter followed by a combination of letters (R, L,
  U, D) and numbers (1, 2, 3, ..., or even 0.5, 1.5, ...) shows a single
  move of the indicated piece. The letters R, L, U, D mean right,
  left, up, and down respectively. The numbers show either some
  part of the size of a single piece (for the Flexible Frame puzzle—
  A), or the number of small unit squares on the bottom of a puzzle
  tray (for the Sliding Weave puzzle and the Beetle puzzle).

6U4, 5D1R4, 8L1D5, 1D5, 2L5,
3L5U1, 4U5, 7D1, 6R6, 1R5U2,
8U4, 5U4, 7L10, 5R2D4R4, 7R4,
8D4, 1D2L5, 6L6, 5U1, 4D5,
3D1R5, 2R5, 1U5, 8U5R1, 7L4U1,
6D4

Figure 5.   Sliding Weave solution (26 moves).

7, 6, 5, 4, 3, 7, 6, 8, 1, 6, 4, 5, 8, 4, 5, 3, 7,
2, 6, 1, 4, 5. Now make a complex move shown
below by the arrow moving pieces $\{8, 3, 7, 2, 1, 5\}$
simultaneously.

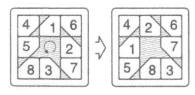
7, 6, 2, 4, 1, 7, 8,
5, 7, 8, 6, 3, 5, 6,
4, 2, 3, 4.

Figure 6.   The Fan Puzzle solution (46 moves; or 41 if the complex move is
considered as one move).

2R1.5, 1R1.5, 3, 5, R, Y, 1, 2, 4,
6, Y, 1, 2, 3, 5, R, 1, 2, R, 5, 3, R,
5R2D1, 3, R, 6L1U1L1U1L1U1, 4,
6, R, 3, 5U1L2, 4, 6, R, 4, 2, 1, 5,
2, 1, Y, 6, R, 4, 1, 2, 3, 1, 2, R, 4,
2R1.5, 1R1.5

Figure 7.   The Beetle solution (53 moves). Solution by Bernhard Wiezorke.

5, 8, 7, 5, 4, 1, 2, 6, 8, 7, 5, 4,
6, 8, 7, 6, 1, 2, 8, 7, 6, 5, 4, 1,
2R0.5D0.5, 7D0.5L0.5, 6, 5, 4, 1,
2R0.5D0.5, 7D0.5R0.5, 8, 6, 5, 4,
1, 2, 3, 8, 6, 5, 4, 1, 2, 3, 8, 6, 5,
4, 1, 2, 3, 8, 7, 5, 4, 1, 2, 3, 5.

Figure 8.   Solution for the Flexible Frame Puzzle—A (61 moves).

Inflated Pentominoes

Rodolfo Kurchan

In October of 1994, in the first issue of my magazine, *Puzzle Fun*, I presented the "Inflated Pentominoes." I believe this is a very rich topic, still open to investigation.[1]

A complete set of 12 pentominoes has a total area of 60 squares and it is possible to make 4 different rectangles, the 3×20, 4×15, 5×12 and 6×10. For example, here is a 6×10 rectangle with the 12 pentominoes.

If some of the pentominoes are inflated, more rectangles are possible. In each of these problems, your goal is to use each of the 12 pentominoes exactly once. (You are free to rotate the pentominoes or flip them over.)

In a *double* inflated pentomino, each square of the pentomino becomes a 2×2 square. For a *triple*, each square becomes a 3×3 square and so on.

1. **One Double** With one double pentomino and 11 single pentominoes we have an area of 75 squares and we can make some 5×15 rectangles. Here is one example found by Brian Barwell.

See if you can find others.

Rodolfo Kurchan is the author of *Mesmerizing Math Puzzles*, Sterling Publishing (2001).

[1]The reader will find a large collection of polyomino and other puzzles in issues of my magazine published on-line [Kur].

2. **More Doubles** Maarten Bos found by computer all the different possible solutions for rectangles using single and double pentominoes.

   There is only one solution using 2 double and 10 single pentominoes in a 5×18 rectangle. See if you can find it. (See solution on page 228.) Hint: You must use these two double pentominoes.

3. **Single and Triple** It is also possible to find solutions using only single and triple pentominoes. See if you can find one. (A 10×10 solution using one triple pentomino is found on page 228.)

4. **Double and Triple** It is not easy to find solutions using only double and triple pentominoes. See if you can find a 15×21 rectangle using the three triples below. (Solution page 229.)

   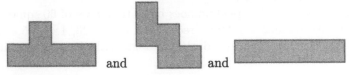

5. **Single, Double and Triple** See if you can find a rectangle using single, double and triple pentominoes. (Two 10×13 rectangles, one 15×15 and one 15×30 rectangle are shown on page 229)

6. **Different sizes** The goal here is to make a rectangle using the 12 pentominoes using as many different scaled factors as possible.

   Michael Reid found a 4-level solution using 1 single, 2 doubles, 8 triples and 1 order 6. I was able to modify his solution to make a 5 level solution using 1 single, 2 doubles, 3 triples, 5 order 6 and 1 order 12. It's an open question whether one can do better. (These two solutions are found on page 230.)

Solutions

The 5×18 rectangle due to Maarten Bos.

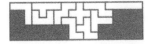

A 10×10 rectangle using one triple pentomino due to Jaime Poniachik.

Here is a solution using 9 doubles and 3 triples in a 15×21 rectangle by Federico Di Francesco.

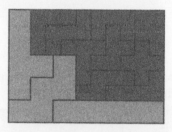

A 10×13 rectangle using 9 singles, 2 doubles and 1 triple pentomino.

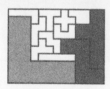

Another solution due to Pieter Trobijn.

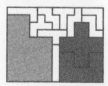

Here is a 15×15 rectangle using 6 singles, 3 doubles and 3 triples.

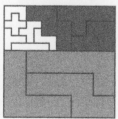

This solution using 1 single, 2 doubles, and 9 triples pentominoes in a 15×30 rectangle was found by Michael Reid.

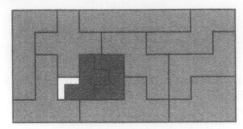

Michael Reid found a rectangle using four sizes of pentominoes, 1 single, 2 doubles, 8 triples and 1 order 6.

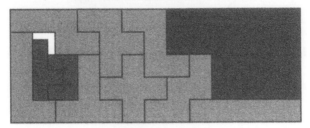

I modified Reid's solution to obtain five sizes of pentominoes, 1 single, 2 doubles, 3 triples, 5 order 6 and 1 order 12.

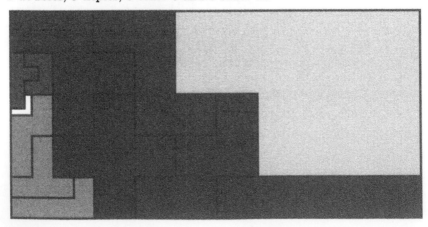

Bibliography

[Gol65] Solomon Golomb. *Polyominoes*. Scribner's, New York, 1965.

[Gol94] Solomon Golomb. *Polyominoes: Puzzles, Patterns, Problems, and Packings*. Princeton University Press, Princeton, New Jersey, 2nd edition, 1994.

[Kur] Rodolfo Kurchan. *Puzzle Fun*. Bulletin. Issues 1–5 and the extra of August 1995 by Maarten Bos discuss inflated pentominoes. http://www.bigfoot.com/~velucchi/pfun/pfun.html.

Pixel Polyominoes

Kate Jones

Martin Gardner's articles on Solomon Golomb's work with pentominoes led directly to my involvement with this form of mathematical recreation. Over 18 years, my company has evolved a product line of over 80 game and puzzle sets primarily of the polyform and combinatorial variety. It will come as no surprise that a large percentage of them deal in one way or another with polyominoes.

In the course of marketing such playthings, I find it interesting that so few people *out there* had ever heard of or seen such puzzles but always and immediately find them fascinating. (That's what keeps us in business.) A great many more people, on the other hand, have or use computers, and until recently I was not among them. In designing our company's web-catalog, I've had to learn to use HTML to mark up text for web pages. I have depended upon my background as a graphic artist to help visualize the finished look, and my 18 years of strenuous puzzle solving carried over to the logic of fitting pieces of totally confusing code together. Among the tasks was the creation of graphics-drawings of puzzle tilings with precise geometric proportions and installed color regions.

It was my good fortune to have the use of a Paint Shop Pro program for this purpose, and one of its means for precise work was an enlargement tool that could get a really good close up, and closer, and closer, and closer ... it could take the worker right into the center of its Universe, down to the fundamental building block of all its imagery ... the unit element of visual illusion ... the mighty pixel. There they were, in the neat rows and columns of a square lattice. And some cells were filled,

Kate Jones is president of Kadon Enterprises, Inc., award-winning publishers since 1980 of over 80 original combinatorial games and puzzles, including two games by Martin Gardner: The Game of Solomon and Lewis Carroll's Chess Wordgame.

and some cells were empty, and from a very great distance the whole composition could be a picture or a letter or even just a line. The diagonal lines were especially intriguing, since pixels are squares and so look like stairs. And there, among the seemingly random groupings of black pixels and empty spaces, it became as clear as only so gigantic a magnifying glass could reveal: pixels in reality form itsy bitsy polyominoes, and polyominoes are just tiny clusters of giant pixels.

To color in the illustrations of the puzzle sets for our Web pages, I used Paint Shop Pro's neat paint can system of "pouring" paint onto a region to be colored. The catch is, the region must be totally enclosed, or the paint will run out all over the place. The trick, then, was to patch any holes or crevices with strategically placed pixels. Just touching at tips of corners is enough to close a gap, so I made a game of trying to see how many different shapes of the pentominoes and hexominoes I could hide in the strewn-about patches of little square black spots. This was purely a fancy of mine, since no one would ever see them at regular size. And this drawing is now on the World Wide Web, at the Kadon site named simply *gamepuzzles*. Everyone on the planet who visits our site and views the pictures will be looking and not seeing the detail of the polyominoes hiding among the pixels. But here is a close-up of what some of the pieces spell in honor of the man whose writings decades ago triggered a series of events that shaped my life.

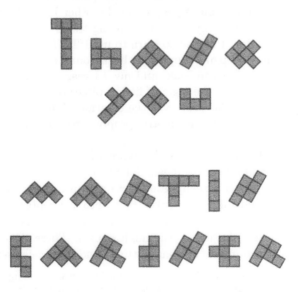

Pythagorean Puzzle

Harry Nelson

Background

Although the statement of the Pythagorean theorem appeared on a Babylonian tablet at least a millennium before the Greek mathematician Pythagoras was born, Pythagoras (fifth century B.C.) is credited with the proof of the assertion that for all triangles the following relationship was true:

> *A square, whose edge was the length of the longest triangle side, had an area which was exactly equal to the sum of the areas of the two squares, whose edges were the lengths of the other two triangle sides, if and only if the triangle was a right triangle.*

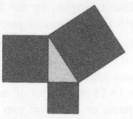

Figure 1. The Pythagorean Theorem: The areas of the small squares add up to the area of the large square.

Harry Nelson, former editor of the *Journal of Recreational Mathematics*, is now a full-time game and puzzle designer and entrepreneur.

A familiar translation of his statement is, "The square on the hypotenuse of a right triangle is equal to the sum of the squares on the other two sides."

As an example, one can *partially* fill up the large square space below with materials consisting of four right triangular tiles in the two different ways shown.

Since the outer squares are both squares of the same area, the inner square area in the left picture must equal the sum of the square areas on the right, providing a geometric proof of the Pythagorean Theorem.

The Puzzle

The puzzle is based on this idea.[1] Construct a 15×15 frame, one 8×8 square block, one 3×3 square block, and four right triangles with legs of length 3 and 8. Start from the center diagram shown below and move the blocks provided into the form of the other diagram at right. You are only allowed to slide the blocks, without lifting or turning them over, and you must always stay within the confines of the outer box. In the right diagram, you'll find that the square does not fit snugly, since the square area surrounded by the four triangles, by the Pythagorean theorem is the sum of the area of both the two square pieces.

Several increasingly challenging versions of the puzzle are suggested. First, slide within the entire 15×15 box space. Second, using the 1×15 strip provided, block off one edge, leaving only a 14×15 working space. Third, inserting the other 1×14 strip, shut off more of the edge, leaving only a 13×15 working space, or block off two edges leaving a 14×14 working space. To solve each of these versions in turn requires the discovery of an increasingly sophisticated set of moves.

---

[1] A handmade wood version of this puzzle is available from the author for $30. He can be contacted at hlnel@flash.net.

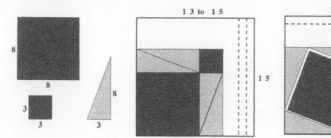
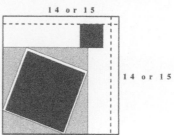

Classic Six-Piece Burr Puzzle

Robert J. Lang

This model was inspired by the burr puzzles of Allan Boardman and Bill Cutler.

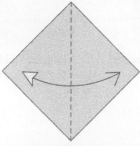

1: Begin with the white side up. Fold and unfold along the diagonals.

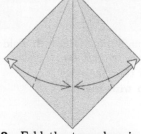

2: Fold the top edges in to the center line, making a sharp pinch along the lower edges. Unfold.

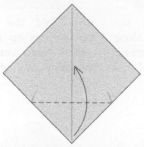

3: Fold the bottom corner up along a line that connects the two pinch marks.

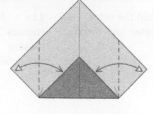

4: Fold the side corners in so their edges line up with the edges of the bottom corner and unfold.

Robert J. Lang is a physicist and engineer but prefers to fold paper. He has written six books on origami.

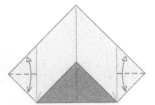

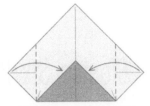

5: Fold angle bisectors from each corner that stop on the vertical creases.

6: Fold the sides in again on the existing creases.

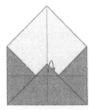

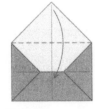

7: Mountain-fold the tip of the corner behind so that the edge lines up with the side corners.

8: Fold the top corner down.

9: Mountain-fold the tip of the corner underneath.

10: Fold the sides in to the center and unfold.

11: Fold the top corners down along creases that connect the midpoints of the sides.

12: Fold the bottom corners up along creases that connect the midpoints of the sides.

13: Turn the model over from side to side.

14: Fold the corners to the center on existing creases, crease firmly, and unfold.

15: Turn the model back over.

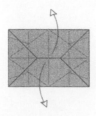

16: Unfold the four flaps.

17: Unfold the top and bottom flaps.

18: Fold four edges in.

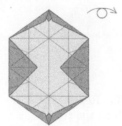

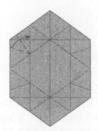

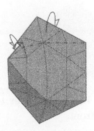

19: Turn the paper over.

20: Form a pleat in the upper left corner. The mountain fold already exists; bring it to the vertical crease. The model will not lie flat.

21: Mountain-fold the edge behind, forming new creases through the pleat.

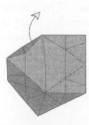

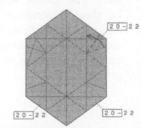

22: Unfold to step 20.

23: Repeat steps 20–22 in three places.

24: Turn the paper over again.

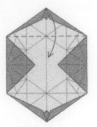

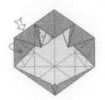

25: Fold the top corner down.

26: Reverse-fold the corners on existing creases.

27: Pleat the corner, using the creases you made in step 20. The model will not lie flat.

28: Bring one layer to the front.

29: Mountain-fold the edge and the pleat underneath, using the creases made in step 21.

30: Repeat steps 27–29 on the right.

31: Repeat steps 25–30 on the bottom.

32: The model will be three dimensional. Make all the indicated mountain folds sharp and push the sides together.

33: Make six such pieces. The puzzle is to assemble them into the 12-pointed geometric solid shown (see Figure 1).

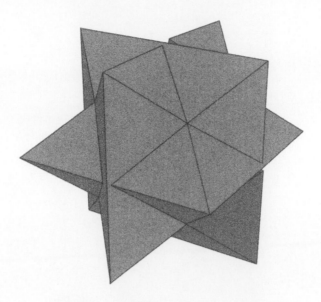

Figure 1.   12-pointed geometric solid.

A Classification of Burrs

Bill Cutler

(See Color Plate XII.)

The term *burr* is used to describe an interlocking 3-dimensional puzzle composed of notched rods of wood. When I was 11 years old, I saw a 6-piece burr in a drugstore window near my school bus stop. I was intrigued by the puzzle, and wanted to learn more about it.

3-Axis Rectilinear Burrs

The most common types of burrs are those which have rods going in three orthogonal directions, and the assembled burr and pieces can be modeled completely by using a 3-dimensional grid of cubes. The rods

---

Bill Cutler is a puzzle designer and computer programmer who frequently combines these activities.

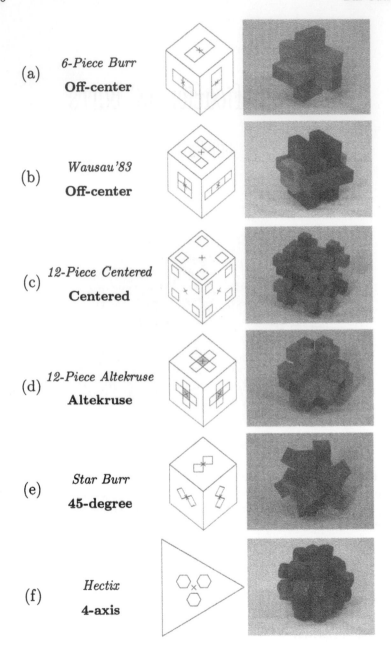

Figure 1. Some examples of burrs. Each burr's classification is shown in bold. The 4-axis burr (f) is tetrahedral; only one face of its four faces is shown in its pattern diagram.

can be of different dimensions, but in many cases, they have a 2×2 cross-section. However, even for rectilinear burrs which use 2×2 square rods, there are several distinct types.

Off-Center Burrs

The 6-piece burr is the simplest version of what I call the *off-center* burrs. In these burrs, all pairs of intersecting rods intersect in one-half the width of a rod. The volume of intersection consists of 4 cubes, each $1/2$ the rod width in size. Either rod can be notched around the other rod, so that the notched rod remains connected, and the un-notched rod can slide back and forth freely. Alternatively, one can arrange the assignment of the 4 cubes in the intersection area to both of the pieces, or to empty space, and so a variety of restrictions on the movements of the rods can be implemented. Therefore, this type of intersection allows for great flexibility in the design of the burr.

Puzzle books contain examples of many different shapes for burrs which fall into the *off-center* category. Two common examples are: (1) the 18-piece burr, which has 6 rods in a 2×3 cluster going in each of the 3 directions; and (2) the 24-piece burr, which features 8 parallel rods in 4 pairs going in each direction, resulting in a construction that has 8 6-piece burrs at the corners of a cube.

Patterns

I visualize 3-axis rectilinear burrs by looking at the arrangements of the rods going in each direction. I draw cross-sections of the 3 sets of parallel rods on the sides of a cube, and call these the *patterns* of the burr. Figure 1(a) shows the patterns for the 6-piece burr.

The patterns for the 6-piece burr are the same in all 3 directions and have 4-fold symmetry about the center. This results in the 6-piece burr having 24 symmetries: 12 rotational symmetries and 12 reflective symmetries. This is true of many other off-center burrs.

Figure 1(b) shows an example of an off-center burr which has less symmetry then the 6-piece burr. The 3 patterns are all different, although they do have 4-fold symmetry about their centers. The resulting burr has only 8 symmetries.

Centered Burrs

Some rectilinear burrs which use 2×2 rods include intersections which are not 1/2 the rod width, but rather are the full width of the rod. Such intersections are more restrictive to include in a burr design—neither rod can be notched around the other without being cut in two. The only option is to notch both rods, in which case the rods cannot slide along their long axis, but must first move away from each other. Figure 1(c) is an example of a *centered* burr.

The *centered* burrs are a class of 3-axis rectilinear square rod burrs in which:

1. all rod intersections are the full width of the rods, and

2. in at least one case (frequently at the center of the burr), rods from all 3 directions intersect at the same place.

The movements are so restrictive in these burrs that in many of them the first move of the puzzle is to twist one or more of the rods, which have had part of a square cross-section trimmed to a circular cross-section.

Altekruse Burrs

The final class of 3-axis square-rod rectilinear burrs is the *Altekruse* burrs, such as in Figure 1(d). In this class,

1. all rod intersections are the full width of the rods, and

2. there are no cases in which rods from all 3 directions intersect at the same place.

The fascinating feature about the Altekruse burrs is the way in which they can be disassembled: The first move involves moving about half of the pieces in one direction, and the other pieces in the opposite direction. After this initial move has taken place, and the pieces are separated enough so that some of the notches no longer fully interlock, then some pieces can be removed in different directions. In many examples of the Altekruse burrs, all of the pieces are notched exactly the same.

Other 3-Axis Burrs

There are also examples of 3-axis square-rod rectilinear burrs which do fall into any of these categories—some of the notches may be full width, and some 1/2 width. These burrs would be placed into a *mixed* or *other* category within the larger class of rectilinear square-rod burrs.

There is a type of burr puzzle which uses square rods running in 3 orthogonal directions, but does not follow the rectilinear structure outlined previously. In these puzzles, which I call *45-degree square rod puzzles*, each rod is rotated 45 degrees around its long axis. The most basic burr in this category is called the *Star Burr*, and has 6 pieces (Figure 1(e)). In many models of this burr, the ends are cut off at an angle, resulting in a shape which has more external symmetry then the underlying burr, and thus hides the true nature of the puzzle. The *Star Burr* is curious in that the six pieces can be notched identically, each with 2 notches, and it seems at first glance that they cannot be put together because there is no *key* piece.

4-Axis Hectix Burrs

The burr types we have seen so far all have rods going in 3 orthogonal directions. These are the directions perpendicular to the sides of a cube. Are there different sets of axes which could be used to create other burr shapes? What if we use one of the other regular polyhedra? The regular tetrahedron has 4 sides, and the perpendiculars to these are 4 axes. These axes are the same as the directions perpendicular to the sides of a regular octahedra, and are also the same as the 4 main diagonals of a cube. What kind of burr shape can be produced using them? Looking down one of these axes, the other 3 axes cross the viewing plane making 60-degree angles with each other. This points towards using triangular or hexagonal rods. I envisioned using 3 hexagonal rods as in the pattern shown in Figure 1(f). This pattern can be used consistently on all 8 sides of the octahedron. When arranged in this way, the rods which intersect will do so in 1/2 of their rod width. The resulting burr consists of 12 hexagonal rods and was discovered independently in the mid-1960s by myself and Stewart Coffin, who was awarded a patent for this design. The center of the burr is a hollow shape bounded by 12 rhombuses— the rhombic dodecahedron—and Stewart thought up Hectix by building outward from this shape. I find it fascinating that we used radically different approaches to arrive at the same *natural* burr shape—Stewart from the inside-out, and myself from the outside-in.

There is a very curious thing to notice about the pieces in Hectix. Pick any of the 12 pieces. This piece intersects 4 other pieces in the puzzle, each of which intersects it halfway through its width. 2 of these pieces run in one direction, and the other 2 pieces are in different directions. Imagine that the first piece is notched to allow for each of the 4 intersecting rods to pass through it uninterrupted. What is left of the rod? Surprisingly, the ends are not detached, but are connected by a fragile helical structure.

6-Axis and 10-Axis

(a) Square-Rod Dodecaplex.   (b) Spider's Web.

Figure 2.

But why stop at 4 axes? There are still 2 Platonic solids left. Take the regular dodecahedron, which has six axes perpendicular to the sides. Looking down each axis, the other 5 axes spread out in an evenly arranged pattern. The natural-shaped rod cross-section would be a pentagon. My 6-axis burr creation, Square-Rod Dodecaplex shown in Figure 2(a), uses square rods rather then pentagonal rods as they are easier to make. The final Platonic solid is the icosahedron (20 sides). The number of axes perpendicular to the sides is 10. The burr I designed of this type called Spider's Web (Figure 2(b)) looks very similar to the Square-Rod Dodecaplex.

Mixture of Axis Types

Having burrs with a mixture of some of these types is possible. The Hybrid in Figure 3 interlaces a 3-axis rectilinear burr with a 4-axis burr. The 3-axis burr is an 18-piece burr with rods separated just enough so that they do not intersect each other; and the 4-axis burr is a 12-piece Hectix with rods similarly separated. The result is an interesting mixture of square and hexagonal rods in which the rods intersect only with their opposite kind.

Figure 3.   Hybrid.

Burr Classification

I. 3-Axis (Orthogonal)
  A. Rectilinear 2x2 Square-Rod
    1) Centered
    2) Altekruse
    3) Off-Center
    4) Mixed
  B. Rectilinear Other
  C. 45-Degree Square-Rod
  D. Other

II. 4-Axis
III. 6-Axis
IV. 10-Axis
V. Mixture of Axis Types
VI. Other

Bibliography

[Cut86] Bill Cutler. Holey 6-Piece Burr! Available from http://www.billcutlerpuzzles.com, 1986.

[Cut94] Bill Cutler. A Computer Analysis of All 6-Piece Burrs. Available from http://www.billcutlerpuzzles.com, 1994.

[Cut78] Bill Cutler. The six-piece burr. *Journal of Recreational Mathematics*, 10(4):241–250, 1977-78.

[Gar78] Martin Gardner. Mathematical games. *Scientific American*, pages 14–26, January 1978.

Paving Mazes

Adrian Fisher

We've been using clay paving bricks in landscape mazes since 1975. England's unique geology provides a remarkable range of different clays, with over 40 different natural clay colors. However, the small range of regular stock shapes significantly limit the possible shapes for a maze.

We found that it was possible to make beautiful and wonderfully detailed images and shapes using a surprisingly irregular paving tile in combination with a square one. Before delving into the advantages of this new and unusual paving system, let's explore the limitations of older paving systems.

Early Paving Projects

Three of our first mazes are shown in Color Plates XIII–XV. While making the Lion Rampant Maze, the noise and dust from cutting bricks with diamond-tipped wheels was unpopular with the local shopkeepers. The street was closed for several weeks while the concrete foundation was cast, and the painstaking bricklaying work carried out.

We tried using the *flexible method*, laying bricks close-butted without mortar on a compact sand base. It is a proven method, much quicker to install, but still places limitation on decorative designs.

During these early maze construction projects, we discovered that while paving bricks could be used to make some intricate designs, we were forced to either stick to patterns with mostly right angles, or

Adrian Fisher of Adrian Fisher Maze Design in England is the leading designer of full-size landscape mazes worldwide. His remarkably varied designs include hedge mazes, tile mazes, maize mazes, and water mazes. The company web site is http://www.mazemaker.com.

to embark on time consuming, labor intensive cutting and bricklaying processes.

In the hopes of overcoming the limitations of off-the-shelf rectangular bricks, we began experimenting with other shapes. Regular hexagonal bricks are available, but they are also too regimented, and therefore lack design vitality.

We tried a pattern using 7-sided and 5-sided shapes, where the smaller piece was a regular pentagon, and all sides were of equal length. We thought this was a marked improvement over rectangular tiles. However in two respects, we were not entirely satisfied; the initial tessellation did not lend itself to achieving straight edges, and it also had a "grain" in one direction (although you hardly notice it in the Oran Utang design).

A New Paving System

We developed a second regular polygonal tessellation, this time with 7-sided and 4-sided pieces, with 5-sided edging shapes, shown in the left image in Color Plate XVI.[1] We've already used this new paving system in half a dozen mazes throughout the world, including The Mall of Georgia in Atlanta shown in the right image in Color Plate XVI.

The new paver system has the following attributes:

1. Design Vitality—the opportunity to create seamless patterns, curves and spirals of any radius, including expanding and contracting spirals, on a close-up human scale (a few meters/several feet across). Square and hexagonal tessellations do not fully achieve this attribute, except on a very large scale.

2. Design "Grit"—the tessellation has an inherent "vibrancy" about it, to encourage the eye to perceive a possible meaning in every side of every piece. These first two objectives allow the artistic "added value" that helps customers justify paying a somewhat premium price.

3. Minimizes the number of different stock items to form a tessellation. Just three shapes are needed.

4. Avoids acute angles, since these chip off in use.

5. Straight sides to all pieces throughout.

---

[1] This has been patented internationally as the Fisher Paver.

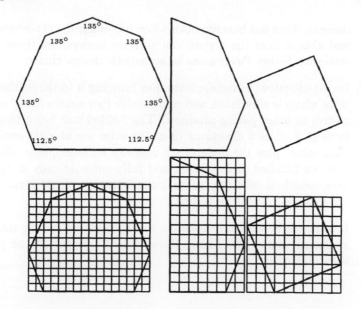

Figure 1. The Fisher Pavers. (Top row) The 5-sided brick is made by cutting the 7-sided brick in half. The third paving brick is square. (Bottom row) Bringing the pavers to market required a few minor adjustments to the shapes. It was advantageous to devise a geometric approximation, with small integer coordinates, that is easy for landscape designers to use on computer.

6. Has no "fault lines" through the tessellation in any direction, which would weaken the stability of the pavement, and have less artistic appeal.

7. Achieves straight edges at the perimeter of the paved area, simply and elegantly, right off the pallet with no cutting.

8. Achieves relatively consistent "path widths" and "barrier widths" when being used to create mazes and other linear patterns just one paver wide. The lengths of the various sides of each paver are not too great or small in proportion to one another.

9. Integrates seamlessly—without cutting—with the modules of other paving systems, in particular (a) the regular 8×4 inch (200×100mm) paving brick, often laid in a herringbone pattern, (b) the various concrete slab systems, based on multiples of 6 inches (150mm), and (c) the Small Square unit (3.2×3.2 inch / 80×80mm) of the Fisher Paver system (thus producing small 5×5 chequerboards within the

design). This has benefits both when adjoining the other systems, and also so that the system can actively incorporate these other materials Fisher Paver areas as an artistic design choice.

10. Is cost-effective to manufacture, thus bringing it to the market at a price which is affordable, and comparable (per square foot / square metre) to other paving products. The 7-sided unit has orthogonal faces in 5 of its 6 directions (3 of the sides are at right angles to each other, plus the underneath and top surfaces), thus allowing it to be handled mechanically and fully-automatically in various axes, which is how clay paver manufacturing plants manipulate the extruded clay prior to firing.

11. In practice is not labor-intensive to lay, thus controlling total installation costs. The Fisher Paver is laid straight off the pallet, with no cutting.

Early Japanese Export Puzzles: 1860s to 1960s

Jerry Slocum and Rik van Grol

Introduction

Japanese export puzzles have introduced millions of people to trick opening boxes, banks and interlocking puzzles. Trick boxes, beautifully veneered with a wood mosaic called "yoseki," require sliding panels to be moved in sequence until the top or bottom lid can slide open. Some boxes require over fifty moves to open. Japanese interlocking puzzles are mostly charming figural shapes, such as a barrel, dog or elephant, that consist of ten to twelve interlocking wooden pieces. Representative samples of mostly old puzzles will be described in this article. This means that, unfortunately, the wonderful modern puzzles designed and made by NOB Yoshigahara (Rush Hour, etc.), Akio Kamei (secret opening puzzles) and other current generation Japanese puzzle designers and makers will not be included. The latter group of puzzles is, however, more easily available to the average puzzle collector than the old and rare Japanese puzzles.

This article was originally presented by Jerry Slocum as an invited lecture at the 18th International Puzzle Party in Tokyo, Japan, in August 1998. Rik van Grol and Jerry adapted the lecture for an article in the October 1999 *Cubism For Fun 50*, published by the Dutch Cubists Club (NKC).

Jerry Slocum is the author of seven books about mechanical puzzles and is also known for his large collection of puzzles and puzzle books. **Rik van Grol** is a Dutch puzzle collector, analyzer, and designer, and is also an editor and publisher of the newsletter *Cubism for Fun.*

Figure 1.   The oldest "Japanese Puzzle" in the Slocum collection is dated 1872, but it was not designed or made in Japan. (See Color Plate IV.)

"Japanese Puzzles"

During my research on the history of Japanese export puzzles, I found that numerous so-called "Japanese Puzzles" are not from Japan. For example, the oldest titled "Japanese Puzzle" in my collection is the beautiful puzzle shown in Figure 1.

Although the puzzle in Figure 1 is named "The Japanese Puzzle," the design was described in a French book, *Les Amusemens*, in 1749 and it was manufactured by E. A. Howland in Worcester, Massachusetts. Why was it called "The Japanese Puzzle"? Probably the title was selected because Japan was closed to the outside world from the mid-seventeenth century until the mid-nineteenth century and there was enormous interest in things from exotic and unknown Japan. So the title helped sell the puzzle.

Puzzles from other countries were dubbed "Japanese" in *Mr. Bland's Illustrated Catalogue of Extraordinary and Superior Conjuring Tricks, etc.*, published in 1889; *Mysto's Magic, Tricks, Jokes, Puzzles Etc.* catalogue of 1911; *C.J. Felsman's Catalogue* of 1915; and the *Scientific Novelty Co. Catalogue of 1930*.

Japan is not the only country's name to be used incorrectly in puzzle titles. We know of "Chinese Puzzles" not from China and "American Puzzles" that have nothing to do with America as well.

Dating Old Japanese Puzzles

Dating Japanese puzzles can sometimes be helped by markings on the puzzle. The McKinley Tariff Act of 1891 required the country of origin to be marked on items imported to the USA. From March 1891 until September 1921, Japanese goods were supposed to be marked with the country of origin. For some reason the Japanese chose to use the word "Nippon" for their marking.

The Act was strengthened in 1921 so that products were supposed to be marked "Japan" or "Made in Japan." From 1945 until 1952 the required marking was "Made in occupied Japan."

After 1952 "Made in Japan" was supposed to be marked on goods exported to the U.S. Much more accurate dating of puzzles can be done by the use of novelty, puzzle and magic catalogues.

The Jeep shown in Figure 2 was made during the 1945–1952 post-war occupation period. The solution sheet is marked but the puzzle itself is not marked.

Figure 2.   Jeep, "Made in Occupied Japan" (1945-1952).

Figure 3.   Catel's catalogue of 1785 included 6- and 24-piece burrs.

Interlocking Puzzles

The earliest examples of interlocking puzzles that I have found were made in Europe.

The 6-piece burr, shown in Figure 3, was called The Small Devil's Hoof and the 24-piece burr was called The Large Devil's Hoof in Catel's catalogue of 1785. Recently David Singmaster, a British historian of mathematical recreations, has found an example of a 6-piece burr in a 1733 Spanish book by Pablo Minguet E. Irol.

The Puzzle Apple and The Puzzle Pear, made in Germany, were shown in the British Conjuring catalogue of Milliken & Lawleys in 1873 (Figure 4).

Figure 4.   Three German interlocking puzzles: an apple, a barrel and a pear.

Japanese wooden interlocking puzzles are called "kumiki." Some sources indicate that they may have begun to be made in Japan in the mid-eighteenth century, about the same time that we know burrs were being made in Europe.

Kumiki originated from the carpenters that designed and made ancient wooden shrines and temples in Japan. It was based on the wooden structural locking joints that did not use nails or glue and were designed to allow wooden buildings to withstand earthquakes. According to books on Japanese toys, the development of a "reformed wood plane," the end of internal wars, and the "spirit of pleasure" in the middle of the Edo period (c.1750) led to the first kumiki.

In *Japanese Games and Toys,* writer Ann Grinham says kumiki came from models that were made to teach woodworker's apprentices how to make and fit wood joints without using nails. The book also says that during the Edo period (1616 to 1866) a 6-piece "plate" puzzle was used for teaching in Japanese schools.

Admiral Perry helped to open Japan to worldwide trade in 1854, after the country had been isolated from the rest of the world for almost 200 years.

One of the first Japanese kumiki designers and makers that we know of was Tsunetaro Yamanaka (1874–1954). Two of his first puzzles, the five-story pagoda and the stork puzzle shown in Figure 5, were made in the 1890s.

Figure 5.   Two nineteenth century "kumiki" designed by Tsunetaro Yamanaka.

The second generation of the Yamanaka family, Kazuich, designed and made vehicles. Animals were developed by Hirokichi Yamanaka in the family's third generation. Currently Shigeo Yamanaka is the family's kumiki designer.

Now let us look at some Japanese wooden interlocking puzzles included in novelty and magic catalogues in the U. S. The earliest Japanese puzzles all came disassembled in boxes.

The Mikado Block Puzzle

The Mikado Block Puzzle (a 6-piece burr, shown on the left in Figure 6), was in the 1915 C. J. Felsman Catalogue. "Mikado" was the title used by foreigners for the Emperor of Japan. *The Mikado* was also the name of a British comic operetta by Gilbert & Sullivan that opened in 1885 to instant success. It was so well known and popular in the U.S. that it made everything Japanese popular.

Is the puzzle really Japanese or was the name "Mikado" used to help sell the puzzle? The actual Mikado puzzle from my collection (Figure 6) does not fully answer the question. The box and the words used on it such as, "The puzzle of puzzles," are typical of Japanese boxed puzzles. But it also says "Made by U.N. Co. N.Y.".

The Yamato Block Puzzle

Another Japanese 6-piece burr was titled The Yamato Block Puzzle. The advertisement, shown also in Figure 6, was in the 1920 catalogue of the Magic Shop, Philadelphia. "Made in Japan" is stated on the label

Figure 6.   The Mikado Block Puzzle (left) and The Yamato Block Puzzle (right).

Figure 7.   Puzzles from the Magnotrix catalogue (1936).

of the puzzle. The other writing is exactly the same as on the Mikado puzzle. Therefore it seems likely that The Mikado Puzzle was made in Japan, in spite of the writing on the box saying "Made in N.Y."

From 1926 to 1936 most of the wooden puzzles being sold in the U.S. were made in Germany. The Johnson Smith novelty catalogue included 21 German wood puzzles. In 1924 the Heaney Magic Co. catalogue included 16 German puzzles. In 1926 the Western Puzzle Works provided a choice of 20 German puzzles.

Some Japanese puzzles, however, continued to be sold during the 1920s and early 1930s. For example, the boxed versions of the Aeroplane and Miyako puzzles were included in the 1931 Lyle Douglas catalogue.

In 1936 the Magnotrix catalogue included Japanese versions of seven standard wooden puzzles that were previously only made in Germany (Figure 7). In addition it included three Japanese figural puzzles, the Battleship, Baby Tank and Locomotive. All but two were sold assembled, a very important change for Japanese exporters trying to sell to the American market.

In 1936 the Japanese succeeded in breaking into the U.S. market with a broad range of wooden interlocking puzzles at very attractive

Figure 8.   Early Japanese vehicle puzzles.

prices, 1/3 to 1/6 of the German prices. The Japanese also added more unassembled figural puzzles in boxes such as those in Figure 8.

In 1937 the Japanese captured the entire U. S. wooden puzzle market. Figures 9 and 10 show the actual puzzles from the 1937 Johnson Smith catalogue. Johnson Smith must have bought huge quantities of these puzzles because in 1944 (during World War II) 25 of these puzzles were still being sold, and even in 1948, 13 Japanese puzzles were still included in their catalogue. Japanese puzzles sold in the U.S. included a wide variety of beautifully detailed cars, trains, trolleys, weapons, ships, aeroplanes, rockets, gates, towers, pagodas, and buildings. A sense for the beautiful variety can be gleaned from Figures 13 and 15.

Figure 9.   Japanese puzzles in the 1937 Johnson Smith catalogue.

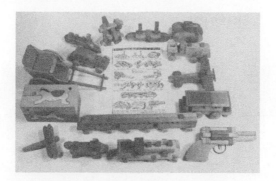

Figure 10.   Japanese puzzles in the 1937 Johnson Smith catalogue.

Figure 11.   Japanese puzzle weapons.

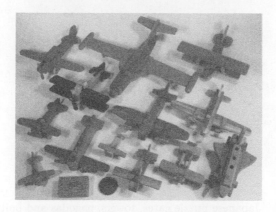

Figure 12.   Japanese puzzle aeroplanes.

Figure 13.   Japanese puzzle rockets.

Figure 14.   Japanese puzzle gates, towers, pagodas and buildings.

Figure 15.   Japanese puzzle animals.

B. Shackman was a large New York novelty company that special-ized in importing novelties and puzzles. Several years after World War II ended, Dan Shackman Jacoby, the grandson of the founder, Bertha Shackman, went to Hakone, Japan and contracted with a co-operative of six small puzzle makers to make copyrighted designs of new puzzles exclusively for Shackman. The B. Shackman catalogue of 1961 included these new Japanese puzzles. Some of the puzzles sold by Shackman are shown in Figures 16 and 17.

Figure 18 shows the *Cash Register* bank. It has "Nippon" stamped on the bottom and it was made between 1891 and 1921.

I have saved the best Japanese interlocking puzzle for last. It is The Tower, shown in Figure 19, by master craftsman Ninomia. Ninomia lives in Hakone and was, and is, Kamei's teacher. Japanese teachers are for life. The Tower has five floors and consists of 106 pieces. It is made of cherry with a walnut base and the doors all open and close. It is

Figure 16.   Some charming and colorful Shackman animals and people. (See Color Plate V.)

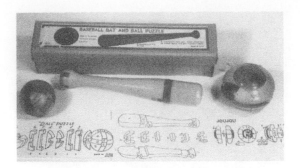

Figure 17.   Shackman's Baseball Bat and Ball and Yo-Yo.

Figure 18.   *Cash Register* bank with "Nippon" stamped on the bottom

Figure 19.   *The Tower* made by Master craftsman Ninomia.

twenty-two inches tall and its grace and beauty are unmatched. Only 10 of these magnificent puzzles were made.

Japanese Puzzle Boxes and Banks

Catel's catalogue of 1785 included a secret opening puzzle box. It is the first known reference to a puzzle box, although some must have existed before 1785.

Within about a decade after Japan began to trade with other countries the Japanese Jewel-Box was sold in the 1867 Adams & Co. of Boston Catalogue. It stated, "Genuine Japanese manufacture".

Jerry Slocum and Rik van Grol

Figure 20.   The Japanese Trick, Match and   **Figure 21.**   A five-book *Puzzle*
Tobacco Box.   *Money Box.*

The Japanese Trick, Match and Tobacco Box shown in Figure 20, was included in the A. Burdette Catalogue from 1877 to 1886. The Smith's Novelty Catalogue also sold it. It appears to be the same puzzle as the Psycho Match-box puzzle in Professor Hoffmann's 1893 book, Puzzles Old and New [Hof93].

A few years later, in 1896, The Martinka & Co. Catalogue showed the Japanese "Inlaid" Puzzle Box (not shown). This box has a drawer which slides out from four different directions. The Johnson Smith catalogues included various money box puzzles from 1926 to 1951, including three sizes of book trick boxes such as the one in Figure 21.

Notes on Japanese puzzle boxes and banks

- Pre-World War II puzzle boxes are made of dark colored wood, are smaller, and frequently have exceptionally fine workmanship.
- Post-World War II puzzle boxes use lighter colored wood and are larger.
- About 100 different designs of Japanese trick boxes and banks are known. They utilize perhaps a dozen types of tricks.
- Some trick banks have concealed coin slots.
- Solutions of puzzle boxes vary from simple—which only require rotating the bottom 90 degrees and removing it—to very tricky and clever solutions requiring numerous steps.

Figure 22. Typical Japanese puzzle boxes from the 1930s to the 1990s. (See Color Plate VI.)

Figure 23. A boat puzzle bank (left) and a water mill puzzle bank (right). (See Color Plate VII.)

Figure 22 shows some typical Japanese puzzle boxes from the 1930s to the 1990s. There are some fine examples of other puzzle boxes in the form of houses, banks, boats and even a water mill, such as those shown in Figures 22 and 23.

Summary

From the variety of early Japanese wooden interlocking puzzles, trick boxes, and trick banks that we have seen, it is clear that the Japanese have made an enormous contribution to the design of interlocking and take apart puzzles.

As far as we know the Japanese began their interlocking puzzle designs about the same time as the European burrs appeared, in the middle of the eighteenth century. More research needs to be done to determine the complete origin and history of interlocking puzzles, but the Japanese figural puzzle designs are original and unique. The Japanese trick banks and trick boxes are also unique. The Japanese introduced many ingenious and very attractive puzzle boxes and puzzle banks to the world.

An even more important Japanese contribution to both interlocking puzzles and trick boxes is the low cost manufacturing methods that they developed. This dramatically reduced the cost of the puzzles, made them affordable, and introduced mechanical puzzles to perhaps millions of households world-wide. Japanese puzzles are still popular all over the world.

On behalf of the community of puzzle collectors, I would like to thank the Japanese for their enormous contribution to the design, and to their innovation in low cost manufacturing, of wonderful mechanical puzzles.

Bibliography

[Gri73] Ann E. Grinham. *Japanese Games and Toys*. Hitachi, Tokyo, 1973.

[Hof93] Professor Hoffmann (Angelo Lewis). *Puzzles Old and New*. Frederick Warne and Co., London and New York, 1893.

[SB86] Jerry Slocum and Jack Botermans. *Puzzles Old & New: How to Make and Solve Them*. Plenary Publications International (De Meern, Netherlands), ADM International (Amsterdam), University of Washington Press (Seattle), 1986.

[Sto92] Tom Stoddard. Still bank collecting—phase ii. *Antique Toy World, May 1992*.

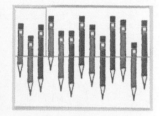

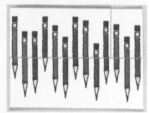

Plate I.   (See page 29.) Mel Stover's "geometrical vanishes."

Plate II.   (See page 4.) Harry Eng's bottles.

Plate III.   Harry Eng and Mel Stover.

Plate IV. (See page 258.) The oldest "Japanese Puzzle" in the Slocum collection is dated 1872, but it was not designed or made in Japan.

Plate V. (See page 267.) Some charming and colorful Shackman animals and people.

Plate VI. (See page 271.) Typical Japanese puzzle boxes from the 1930s to the 1990s.

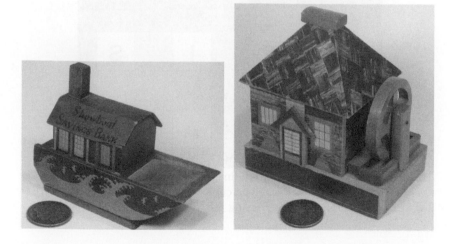

Plate VII. (See page 271.) A boat puzzle bank (left) and a water mill puzzle bank (right).

Starting from   End result

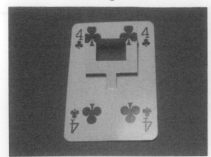 

Plate VIII. (See page 147.) Note how the square frame is folded down, but the leg (or spine) is on the *wrong* side of the frame. One would expect the leg to be *under* the frame rather than *over* it.

Plate IX. (see page 143.) The Multidimensional Gardner (MG). This mystic figure encodes the 4-dimensional hypercube in Plate X. Traversing, say, a RED line in Plate X corresponds to crossing a RED band in the MG.

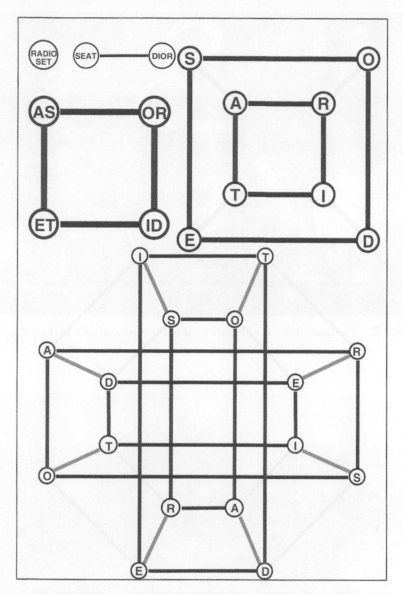

Plate X. (See page 144.) The first five n-dimensional hypercubes: 0-D hypercube or point (RADIO SET), the 1-D hypercube (SEAT-DIOR) line segment, 2-D square (AS-OR-ET-ID), 3-D cube (S-O-D-E-T-I-R-A), and the 4-D tesseract at the bottom.

Plate XI. (See page 209.) A wide variety of puzzles are available to the collector!

Plate XII. (See page 245.) Interlocking burr puzzles.

Plate XIII. (See page 253.) The Lion Rampant Maze.

Plate XIV. (See page 253.) The maze at the New Milton Junior School in Hampshire, England, was installed using no mortar.

Plate XV.   (See page 253.) The Oran Utang Pavement Maze at the Edinburgh Zoo.

Plate XVI.   (See page 254.) Mall of Georgia (Atlanta, GA, USA).

Interlocking Spirals

M. Oskar van Deventer

Three Interlocking Spirals

As part of my never ending search for puzzle mechanisms, I have been looking for ways of putting three disks through each other. I first tried to fit three flat spirals through each other. After quite some calculation, I concluded that the biggest linear spirals that would fit together were 5/12 (41.66%) spiral and 7/12 (58.33%) air. Figure 1 shows how the three interlocking spirals fit together.

Figure 1. Three interlocking spirals.

M. Oskar van Deventer is the creator of hundreds of innovative mechanical puzzle designs, several of which are commercially available.

273

The three spirals have the intriguing property that you can all rotate them simultaneously. The whole constellation does not have much symmetry. It only has a threefold symmetry around one of the principal diagonals of the cube. The three-spiral object turned out not to be very stable, because the spirals touch each other at only a few points. In fact, there is only one way that you can put it down so that it is stable on a flat surface. In any other position, the spirals will move by gravity and the object gets distorted.

To make the object into a puzzle, I have made the spirals 50% spiral and 50% air, and squeezed the spirals into an oval shape. The way in which all three spirals become bent makes them more difficult to put together as a puzzle. However, the structure is still unstable. A more successful puzzle uses three identical linear spirals with 50% spiral and 50% air, with some carefully placed notches in the edges of the spirals. When bringing the spirals into their final position (quite a puzzle!), the notches click together and finally, the puzzle remains stable together.

Four Interlocking Spirals

Continuing my search with the spirals, I tried putting more of them together. To my surprise, I discovered that four spirals can be put together with tetrahedral symmetry. My surprise was even greater when I found that the spirals could be 50% spiral and 50% air and that the spirals touch each other at every critical point! Figure 2 shows how the four interlocking spirals fit together.

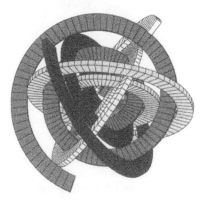

Figure 2.   Four interlocking spirals.

The four spirals also have the curious property that you can rotate all of them simultaneously. When checking the symmetries of the constellation of four spirals, I found that it has the isometry of a tetrahedron. It can actually be easily proven that this is the only possible constellation of spirals that is isometric! The proof is related to the fact that the tetrahedron is the only isometric object that has a corner pointing upward when it lies on the floor. You may try to complete the proof yourself.

The four interlocking spirals form a nice mechanical puzzle. It can be taken apart by pushing it flat. A puzzle made of thick, stiff material may resist. When it is flat, the puzzle can be disassembled by rotating all spirals simultaneously and unscrewing them. If you think that that is difficult, then try to put them together!

You can make this puzzle from cardboard. You will discover that the puzzle is quite material efficient, as you can make two spirals from one piece of cardboard (Figure 3). I had some samples made for me by water-jet cutting polystyrene. This material is flexible and rugged, so it is quite suited for the puzzle. The result is a very colorful red-yellow-blue-green combination, which I have put on a black tetrahedral pedestal.

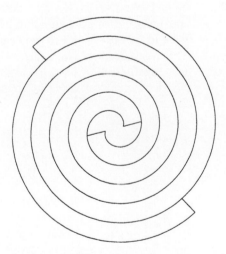

Figure 3. A template to cut out your own spirals.

The Partridge Puzzles

Robert Wainwright

Introduction

As a lifelong participant in mathematical recreations and games, I am
delighted to contribute to this publication honoring Martin Gardner.
Over the years, I have submitted material to his "Mathematical Games"
column regarding various topics including square tiling. One of my problems, *The Partridge Puzzle,* has gone through some fascinating developments during the last two decades. We begin with the basic problem.

Packing a Partridge in a Square Tree

It's well known that the sum of the first i cubes must be a perfect square.
In particular,

$$1 \cdot 1^2 + 2 \cdot 2^2 + 3 \cdot 3^2 + \cdots + i \cdot i^2 = (1 + 2 + 3 + \cdots + i)^2 = N^2,$$

where $N = i(i + 1)/2$. This suggests a tiling problem of efficiently
packing square tiles from the set $(1, 2, 2, 3, 3, 3, \ldots, i, \ldots, i)$ into a large
square of side length N. Observe that regardless of the value of i, the
combined area of the tile set exactly equals the total area of the large
square. Further, the total number of tiles equals the side length of the
large square. Overlapping of tiles with themselves or the border of the
large square is not allowed.

Robert Wainwright is best known for establishing *Lifeline,* a quarterly
newsletter about John Conway's Game of Life.

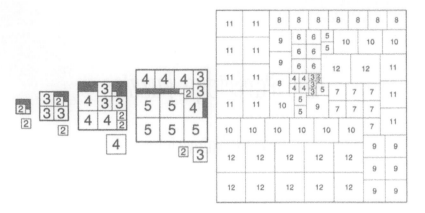

Figure 1.   For $i = 2$, 3, 4, and 5, perfect packings are not possible; one or two tiles must be left out. When $i = 12$, a perfect packing is possible—"A Partridge in a Square Tree."

Our problem concerns covering the maximum area of the larger square using the smaller square tiles. For cases when i is small (excluding the trivial case for $i = 1$), it's impossible to cover the entire $N \times N$ square. Figure 1 shows the best possible packing arrangements for $i = 2$, 3, 4, and 5 as well as the resulting leftover or residual area. As i increases further, this residual area seems to decrease, while the number of packing arrangements to be investigated grows dramatically. In particular, a solution for any odd i (e.g., 7 or 9) can be at least as good as one for $i - 1$ (e.g., 6 or 8) since i squares of size i can always be placed (wrapped) around two adjacent sides of an $(N - i) \times (N - i)$ square to form an $N \times N$ square.

Figure 1 also shows a perfect solution for $i = 12$, a complete packing with no leftover area. An analogy with the verse from the popular twelve days of Christmas song later led to the title of this problem.

These, and larger, solutions were described in my original letter to Gardner. At that time, Gardner informed me he was not aware of any previous work in this area and the problem seemed interesting. He wrote Ronald Graham, an expert on square packing problems, who confirmed that the problem appeared original.

When Gardner was preparing material about packing squares for Chapter 20 of his book, *Fractal Music, Hypercards and More*, he decided to include this problem as a related task. He also gave my solution for $i = 12$, but emphasized the lack of proof for impossibility of any of the smaller results.

In June 1993, Charles H. Jepsen of Grinnell College and Stephen Ahearn, a student, reported some interesting new results, including a perfect solution for $i = 11$. Further, they reported no solutions are possible for $i = 6$ by systematic exhaustion of all possible cases for the size 21 square. At that time, the problem still remained unsolved for $i = 7, 8, 9$, and 10. In addition, they also found a perfect solution for $i = 16$ containing many blocks of smaller squares which in turn yielded solutions for $i = 17$ through $i = 33$.

In January 1996 at the second Gathering for Gardner I presented The Partridge Puzzle along with Jepsen's discoveries.

Packing a Partridge in the Smallest Square Tree

A few weeks after presenting the above results, I received a letter from Bill Cutler indicating he had modified his "BOX" software program to search for smaller solutions to this problem. Through exhaustive search, Cutler indicated that no solution was possible for $i = 7$. Further, he reported a specific solution, the minimum size possible, for $i = 8$ (which leads to a solution for $i = 9$), and later for $i = 10$ and 11. With these discoveries, Cutler put to rest the basic questions regarding low order solutions.[1]

Later that same year, I received correspondence from other individuals who independently discovered minimal solutions. These included William Marshall, Michael Reid, and Nob Yoshigahara. Marshall also found separate solutions for $i = 9$ and 10, the later leading to solutions for $i = 11$ through 21 due to existence of separate blocks of smaller tiles. I presented these discoveries at the third Gathering for Gardner in January 1998.

Packing a Partridge in a Non-Square Tree

Several variations to this basic problem have been examined. Originally I explored other geometric shapes based on the concept of the sum of cubes being equal to the square of their sum. For example covering N by rN rectangles with tiles of sizes up to i by ri tiles. When $r = 2$ a perfect solution exists for $i = 8$. This non-trivial solution, shown in Figure 2(a), was presented to the gathering in Atlanta and inspired

---

[1] I invite the reader to try to pack one 1×1 square, two 2×2 squares, three 3×3 squares, ..., and eight 8×8 squares perfectly in a 36×36 square. For the solutions of this puzzle and others, visit the book's web site at www.g4g4.com.

Cutler to search for perfect solutions with smaller tile sets. Again, using his BOX computer program, he discovered smaller non-trivial solutions including for $i = 7$ with $r = 2$, and $i = 6$ with $r = 3$.

Solutions for different geometric shapes have been investigated by others as well. For example, Marshall reported the perfect solution shown in Figure 2(b) for $i = 9$ using equilateral triangle tiles packed into a larger equilateral triangle shape. He believes this to be the minimal perfect solution. In another case, Reid discovered the perfect solution shown in Figure 2(c) for $i = 4$ using trapezoidal tiles packed into a similar large shape.

Two correspondents, Colin Singleton and Don Knuth took minor issue with my interpretation of the "Twelve Days of Christmas" theme. They pointed out that, although the song is repetitive, it is the unit-item which is mentioned twelve times, and the twelve item only once.

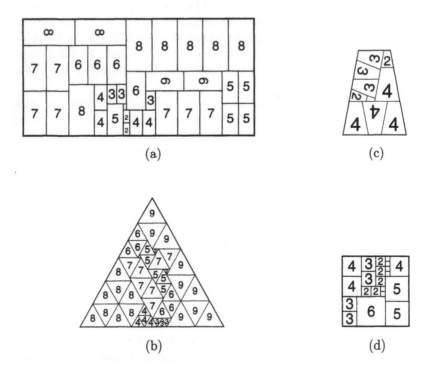

Figure 2.   Solutions to three variants on the partridge problem: (a) A non-trivial rectangular tiling, (b) Marshall's surprise, (c) Reid's surprise, and (d) Singleton's suggested sequence.

Rather than the original equation it was suggested that the following be considered using j rather than i to denote the order:

$$j \cdot 1^2 + (j-1) \cdot 2^2 + (j-2) \cdot 3^2 + \cdots + 1 \cdot j^2 = j(j+1)^2(j+2)/12$$

This is interesting except that all values of j do not result in perfect square sums. Singleton pointed out that such sums do exist for orders $j = 6$, 25, 96, 361, etc. and offered the solution, albeit trivial, for $j = 6$ shown in Figure 2(d). Recently Robert Reid submitted a less trivial solution for $j = 25$ made up of 325 tiles.

Another variation to the partridge problem involves arranging a given set of square tiles into a rectangular rather than square shape. Marshall submitted two such examples using the minimum solution $i = 8$ tile set. Rectangles with dimensions 24×54 and 27×48 can each be created using this same set of tiles.

Most of these extensions were presented at the fourth Gathering for Gardner in February 2000 in a paper titled "Packing a Partridge in a Non-Square Tree."

Part VI
Mathematical Entrees

Fermat's Last Theorem and the Fourth Dimension

James Propp

What's the Problem?

Fermat's Last Theorem has got to be one of the most popular problems in the history of mathematics—millions of people have toyed with it, and thousands have worked up a real mental sweat trying to solve it. The problem, posed by the French mathematician Pierre de Fermat back in the seventeenth century, is usually stated in terms of the famous equation

$$x^n + y^n = z^n,$$

where x, y, z, and n represent unspecified whole numbers. When $n = 1$ the equation has too many solutions to be interesting, and when $n = 2$ there are still infinitely many ($3^2 + 4^2 = 5^2$ is the most famous). The problem Fermat bequeathed to us is to show that when n becomes bigger than 2, the situation changes dramatically: there are no solutions at all. That is: when n is a whole number bigger than 2, no number that is the nth power of a whole number can be written as the sum of two smaller nth powers.

It stands to reason that a proposition so tantalizingly simple would have a simple proof or a simple disproof. Yet for over three centuries the

---

James Propp is a professor of mathematics. He is writing a book on Fermat's Last Theorem for the mathematically interested public. The illustrations were prepared by David Feldman.

problem resisted the efforts of the sharpest minds that tackled it—and we still don't have a simple proof.

Fermat's Last Theorem came to light after Fermat's death, when his son Clement-Samuel was cleaning up the old man's library. An especially cherished work in the elder Fermat's collection had been a seventeenth-century Latin edition of a millennium-old Greek treatise on numbers by the mathematician Diophantus of Alexandria. On one page, Diophantus discussed the problem of writing a given square as a sum of two squares; writing in the margin of that page, Fermat made his no-go claim about higher powers and famously said he'd found a wonderful proof of this result but couldn't include it because the margin was too small.

Fermat as Publicist

The claim is not found elsewhere in Fermat's known writings, but on several occasions he did state that a third power can't be the sum of two smaller third powers, or a fourth power the sum of two fourth powers. However, in the combative fashion of the times, Fermat would often announce his results indirectly, by proposing challenges for other mathematicians to test their wits on. He thought that these challenges would give others a greater appreciation of the hidden depths surrounding his problems about numbers and lure them into doing active research on the topic, but sometimes the tactic backfired on him.

For instance, in one of his letters he challenged the English mathematician John Wallis to solve two problems:

1. given a cube, to write that cube as a sum of two cubes; and

2. given a sum of two cubes, to write that number as a sum of two cubes in a different way.

The first problem has no solution; in fact, this is just the case $n = 3$ of Fermat's Last Theorem. The second problem has many solutions; for instance, $(3/2)^3 + (5/3)^3$ can also be written as $(2)^3 + (1/6)^3$ (Fermat was concerned here with fractions as well as whole numbers). What Fermat seems to have wanted was for Wallis to demonstrate that the first problem had no solutions and then to give a systematic approach to solving the second problem. That is the real two-part challenge Fermat had in mind.

But from the way Fermat wrote the challenge, the first part seems to be asking Wallis to do something that is in fact impossible, and that Wallis probably suspected was impossible (perhaps after hours of fruitless work). It's understandable that Wallis resented this sneaky way of disguising the nature of the challenge, and later missed few opportunities to disparage Fermat's work on numbers.

Although Fermat's efforts to interest his contemporaries in problems about numbers were unsuccessful, he did find followers posthumously, starting in the century after his death. It was up to these disciples to fill in the blanks in Fermat's work, since Fermat himself had been loath to write down details. By the middle of the nineteenth century, all of Fermat's many claims had been proved (or, in a case or two, disproved), with the exception of the famous marginal note. This "Great Theorem of Fermat" also acquired the name "Fermat's Last Theorem" to mark its recalcitrance. Nowadays many people call it "FLT" for short.

What if ...?

Ironically, everyone knew how the proof of FLT should begin: "Suppose there did exist non-zero whole numbers x, y, z satisfying $x^n + y^n = z^n$, with $n > 2$. Then ...." In mathematics, to prove that something doesn't exist, it's frequently helpful to assume for argument's sake that the thing does exist, and then show that the thing, merely by existing, would have to possess mutually incompatible properties, thus demonstrating that it couldn't exist in the first place. This is the method of proof by contradiction, or *reductio ad absurdum*, and it's the method of choice for a problem like this.

So, people knew what the seed of the proof should be, but there has to be some sort of soil into which a seed can be planted. Fermat himself, back in the seventeenth century, seems to have tried planting the seed in the obvious place: the study of the properties of ordinary whole numbers. This study nowadays is called elementary number theory (to distinguish it from the more abstract developments that came later). Leonhard Euler, who as the first of Fermat's posthumous disciples revived the study of numbers in the eighteenth century, was able to construct proofs of FLT for the cases $n = 3$ and $n = 4$ (proofs conceivably found earlier by Fermat), using elementary methods. But going beyond $n = 3$ and $n = 4$ was hard. Euler's successors, and their successors up till the middle of the nineteenth century, were able to settle a few more cases, but this approach petered out and couldn't even be made to handle a value of n as small as 11. It seems that the ground of elementary number theory just doesn't have the right sort of nutrients for the seed of the proof of FLT—the kernel of contradiction—to sprout and grow into a full and rigorous argument.

In the middle of the nineteenth century, mathematicians like Ernst Eduard Kummer found a different plot of land to plant the seed in: a new sub-discipline within number theory called algebraic number theory, and a sub-sub-discipline called the theory of cyclotomic number rings. Cyclotomic number rings are extensions of the ordinary arithmetic of whole numbers, in which other sorts of numbers, including imaginary numbers like the square root of minus one, are brought into the game.

With the new methods, it became possible to prove FLT for many more exponents. Kummer more or less settled FLT for all exponents under 100 (he made a few mistakes on the hard ones). When Kummer's mistakes were corrected and his methods were extended and married with the power of twentieth-century computers, it became possible to prove FLT for all exponents up into the low millions. But, for all mathematicians knew, these corroborations were a fluke; FLT might have been false not just for one exponent, but for infinitely many exponents— perhaps even for *all* prime exponents with more than a million digits, say.

Someday mathematicians might know enough about cyclotomic number rings to be able to construct a proof along the lines that Kummer envisioned; but it seems that the soil of algebraic number theory, in its current state, doesn't have the right nutrients either.

FLT's Last Century

Over the course of much of the twentieth century, professional interest in Fermat's Last Theorem as a hot research topic dwindled. The problem was still part of the lore of mathematics, and part of the field's long-term agenda, but mathematicians found it hard to come up with new plans of attack that hadn't already been tried. No one had an idea how to proceed with FLT, and some experts even began to suspect that Fermat might have guessed wrong.

But outside the academies, more people were working on the problem than ever before. Amateurs were attracted to the problem for a number of reasons. First, FLT is a simple and catchy question. Second, the fact that Fermat claimed to have found a proof raised people's hopes that a proof, indeed a simple proof, could be found. Third, there are certain people who are attracted to a problem precisely because it's hard, and here was a problem that a whole community of experts, the world's mathematicians, had despaired of solving with existing tools. Fourth, there was a cash prize for the person who solved the problem. And fifth, it's easy to *almost* prove Fermat's Last Theorem, in a certain sense.

Remember the basic strategy for proving FLT: you assume that it's false and derive a contradiction. Well, it's very easy to arrive at contradictions in mathematics—just make one mistake and, unless you inadvertently make another mistake that cancels out the first one, you're likely to hit on two assertions that don't square with each other. Even if you find your mistake, or it's pointed out to you, and you realize that your attempted proof by contradiction isn't valid, it's easy to convince yourself that, since you found a proof of FLT with only one mistake in it, you might be close to finding a proof with none. This psychological effect made Fermat's challenge a very addictive problem to work on. But despite the serious efforts of very many people, with various degrees of persistence, no one could find a proof.

Finally, in the last decade of the twentieth century, mathematician Andrew Wiles, aided by his former student Richard Taylor, gave a proof of Fermat's Last Theorem. The proper soil for the seed, or at least one proper soil for it, had been found: an area called the theory of elliptic curves, whose borders Fermat himself had rambled across but whose true shape didn't emerge until the nineteenth century, and whose central inner jungle, still untamed today, is being explored by number-theorists with the aid of powerful ideas from all across the spectrum of mathematics.

And here we face a paradox: Even though Fermat's problem itself is the epitome of popular mathematics—easy to state, ponder, and play with—the eventual solution has the opposite character: it's as esoteric as can be, and it uses ideas from areas of mathematics that didn't even exist in Fermat's day.

It's nonetheless possible to say some things about the proof that are accessible at a popular level. For instance, one key feature of the new approach to Fermat's Last Theorem, and some would say the reason for its success, is the way in which it brings geometry into the story. Not just any old geometry, but the *right kind* of geometry.

The Geometry of the Torus

Picture an ordinary rectangular sheet of paper. If you were to roll up such a piece of paper as shown in the left panel of Figure 1, joining the left edge to the right edge, you would get a cylinder with a vertical axis of rotational symmetry. On the other hand, if you had rolled up the piece of paper as shown in the right panel of the figure, joining the top edge to the bottom edge, you would have gotten a different cylinder, with a horizontal axis of rotational symmetry.

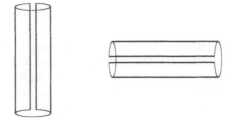

Figure 1.   Two ways of bending a piece of paper to form a cylinder.

Could you have it both ways, joining left to right and top to bottom
and creating two axes of rotational symmetry? The answer is No, if
you limit yourself to three-dimensional space: no matter how you try to
bend the two ends of a cylinder together, you can only create a new axis
of symmetry by destroying the one that was there before.

But suppose that, after turning the two-dimensional rectangle into a
cylinder that bends around in the third dimension, you could somehow
bend that cylinder around through the fourth dimension. Then it turns
out you can get a shape with two axes of full rotational symmetry.
Mathematicians have known about shapes like this for over a century.
But the real surprise was that Fermat's Last Theorem can be understood
as making covert reference to the properties of certain shapes like this.

How can we get a handle on such a shape if we can't build it? If
you're comfortable with the idea that a point in four dimensions can
be specified by four numbers, or in some sense "is" just a quadruple of
numbers, then the shape we're after can be described as the set of all
quadruples of numbers (s, t, u, v) satisfying the two equations $s^2 + t^2 = 1$
and $u^2 + v^2 = 1$. The first equation describes a circle, and no part of
a circle looks any different from any other part. The second equation
also describes a circle. Combine the two circles in a four-dimensional
way and you get a circle-of-circles in which every part looks the same as
every other.

A more geometric way to understand this mysterious symmetrical
shape is to consider distorted versions of it in ordinary three-dimensional
space. If we drop the constraint that there be two axes of rotational
symmetry, and brutally join the two ends of a cylinder in the third
dimension rather than the fourth, we get a shape like the surface of a
doughnut, as shown in Figure 2. (You can't do this with a cylinder made
of paper, but a stretchable surface would work.) With its one axis of
rotational symmetry, this doughnut is only an inadequate shadow of the
more symmetrical shape that lives in four-dimensional space.

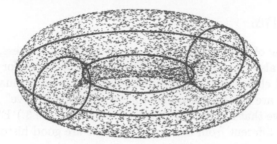

Figure 2.   A torus.

There's a different way to try to understand the symmetrical surface, by ignoring the space that it sits inside and imagining instead what it would be like to be confined to it. Picture an ant on the original rectangle. If the left and right edges of the rectangle are joined, forming a cylinder, the ant's point of view is that when it crosses the join, it's transported from the left edge of the rectangle to the right edge, or vice versa. Now suppose we had a kind of magic paper which, without the need for any bending, would miraculously transport the ant from the left edge to the right edge and vice versa. To us, this scenario would look very different from the non-magic scenario, in which ordinary paper is rolled up into a cylinder; but from the ant's point of view, the two are the same.

Now suppose we had a fancier brand of magic paper which, without any bending, would miraculously transport the ant from the left edge to the right edge and vice versa, *and* from the bottom edge to the top edge and vice versa. From the ant's point of view, there's no difference between the rectangular universe of magic paper that it inhabits and the symmetrical surface in four dimensions that we're trying to understand.

Magic paper doesn't exist, but computers that can simulate it do. There's even money to be made from creating such simulations, as was discovered twenty years ago by the inventors of the popular video arcade game Asteroids. Nowadays, thanks to the World Wide Web, you don't even need a stack of quarters to experience what it would be like to live on such a surface; you can play Mike Hall's web-version [Hal]. Other games situated in the magic-paper universe can be found at the TorusGames site [Key]. With the help of such programs, this magical, un-makable surface, this creature whose natural habitat is the fourth dimension, can be made amenable to (virtual) exploration. (For other four-dimensional fun, try Rich Schwartz's game Lucy and Lily [Sch].)

Elliptic Curves

As the name of the second website might lead you to guess, a surface of this kind is also known as a torus. There's another name for tori: they're often called elliptic curves. This nomenclature is unfortunate, since the uninitiated are apt to think that the term "elliptic curve" refers to the kind of curve that's called an ellipse (shown in Figure 3.) Elliptic curves are totally different from ellipses. But there are good historical reasons for this confusing terminology.

Let's go back and talk about the ellipse a bit. An ellipse is what you get when you view a circle from a slant. The ancient Greeks studied the ellipse, but the shape didn't come into its own until the seventeenth-century astronomer Johannes Kepler discovered that the orbits of the planets are better approximated by ellipses than by circles.

Mathematicians of the seventeenth century, developing ideas that would later become the calculus, tried to find a formula for the circumference of an ellipse, and failed. It turned out that new kinds of mathematical functions had to be invented, just as the ancient Greeks had had to invent the sine and cosine functions in order to solve *their* problems about triangles.

These new functions turned out to be useful for lots of real-world problems: for instance, studying the behavior of swinging pendulums or buckling beams. But because the functions rose to prominence from their use in measuring the circumference of ellipses, they became known as elliptic functions.

Now, if you studied trigonometry, you probably did it twice: the first time trig was about properties of triangles, and the second time it was about properties of the circle of radius 1. In fact, trig functions like sine and cosine are sometimes called "circular functions" to honor the way they relate to properties of the circle. Something analogous happens with elliptic functions: curves like the one shown in Figure 4 give you a

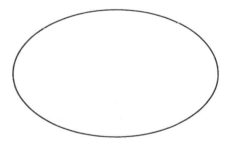

Figure 3.   An ellipse.

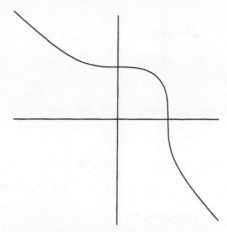

Figure 4. The curve $x^3 + y^3 = 1$.

good way of thinking about elliptic functions. So curves like this became known as elliptic curves.

Just as the circle is given by the equation $x^2 + y^2 = 1$, this particular curve is given by the equation $x^3 + y^3 = 1$. Figure 5 shows another elliptic curve, with the equation $y^2 = -(x^2 - 1)(x^2 - 9)$. Algebraists call the locus one curve, even though it has two components, because it's given by a single equation.

So this is what mathematicians call an elliptic curve (and why)—but how do we get from here to doughnut-shaped surfaces in four-dimensional space?

The answer is: instead of looking at pairs of real numbers x, y that satisfy the algebraic relation $y^2 = -(x^2 - 1)(x^2 - 9)$, look at pairs of *complex* numbers that satisfy that relation, such as the pair $x = 0$ and $y = \sqrt{-9}$.

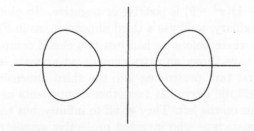

Figure 5. The curve $y^2 = -(x^2 - 1)(x^2 - 9)$.

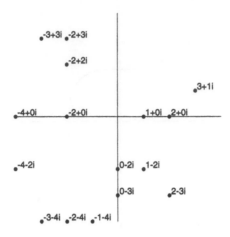

Figure 6.   Some complex numbers plotted in the plane.

Complex Numbers

Every complex number can be written as $a + bi$, where i is the square root of minus one and where a and b are ordinary real numbers, called the real and imaginary parts of that complex number. So we can plot a complex number in two dimensions, by plotting the point (a, b), as shown in Figure 6. (For instance, $\sqrt{-9} = 0 + 3i$ would be represented by the point $(0, 3)$ on the vertical axis.) But to plot a *pair* of complex numbers, or to draw the graph of an equation involving two complex variables, you need two plus two dimensions—that is, you need to draw a surface in a four-dimensional space.

To get a peek at the four-dimensional surface that's latent in our two-dimensional picture Figure 5, let's step halfway into the fourth dimension by stopping at the third. We're going to keep x a real number, but we're going to let y be a real number or an imaginary number, according to whether $-(x^2 - 1)(x^2 - 9)$ is positive or negative. To plot the value of y when it's imaginary, we'll use a third dimension, as in Figure 7.

Notice that where before we had just two closed components of the curve, we now have three: an extra component appears in the middle, touching the first two, protruding into the third dimension. Our new picture of the elliptic curve has two other components as well, one on the right and one on the left. They go off to infinity, but the nineteenth-century mathematicians who invented projective geometry advised us that in contexts like this, it's appropriate to add in extra points at

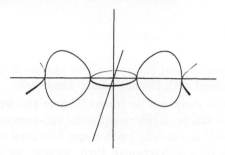

Figure 7. The plot of $y^2 = -(x^2 - 1)(x^2 - 9)$ with x real and y arbitrary.

infinity. If you take their advice, you'll view these two unbounded components of the curve as actually meeting at a point at infinity.

So, schematically, what we get are four closed curves forming a kind of necklace, with each curve touching two of the others. But this is exactly what we have when we draw four circles on the surface of a doughnut, as shown in Figure 2!

So far, we've required x to be a real number. When you let x and y be any old complex numbers satisfying $y^2 = -(x^2 - 1)(x^2 - 9)$, the three-dimensional backdrop of Figure 7 becomes the spine of a larger four-dimensional backdrop, and the four closed curves become a kind of skeleton that the rest of the elliptic curve hangs on. I won't show you the details, but at least I hope you can see that the doughnut shape doesn't come out of nowhere. Moreover, it turns out that when you take this complexified four-dimensional picture and bend it in the right way, the curves of Figure 7 becomes four perfect circles. Likewise, when the curve $x^3 + y^3 = 1$ shown in Figure 5 is complexified and suitably bent, then the original curve, lying inside the surface that arose from it, forms a perfect circle.

This discussion barely scratches the surface of elliptic curves and their symmetries. For instance, consider the seemingly unrelated, recreational problem of placing 12 dots in the plane so that every line that goes through two of them goes through exactly one other point. How big could the number of such lines be? The answer is 19, and if you want to draw such a picture with 12 points and 19 lines, the prettiest way is to choose all sixteen points to lie on the curve given by the algebraic relation $(x - 1)((x + 2)^2 - 3y^2) = 8$. This is an elliptic curve with threefold symmetry, but its four-dimensional unfolding has even more symmetry; in fact, if you lift the curve up into the fourth dimension, so that it becomes a circle on a torus, then the twelve points are evenly spaced! (See the article "Planting Trees" by Stefan Burr [Bur81].)

Back to Fermat

How did these symmetrical surfaces, these visitors from the fourth dimension, get involved with Fermat's Last Theorem?

This part of the story can be traced back to the second half of Fermat's challenge to Wallis: if you've got some non-zero number c that can
be written as $a^3 + b^3$ in one way (with a and b rational numbers—that is
to say, whole numbers or fractions), then, leaving aside the case where
$a = b$ or $a = 0$ or $b = 0$, Fermat said that there must be another way to
write c as $a'^3 + b'^3$ with a' and b' two other rational numbers. That is,
in addition to the point (a, b) and its twin (b, a), there must be another
"rational point" (a', b') on the curve $x^3 + y^3 = c$, where a point (x, y) is
called rational if both of its coordinates x, y are rational numbers.

The general notion of an elliptic curve didn't exist in Fermat's day,
but the curves of the form $x^3 + y^3 = c$, along with other specific curves
studied by Fermat, are in fact examples of elliptic curves. Fermat had
some tricks for finding rational points on his elliptic curves, and these
were developed further by later mathematicians, notably Isaac Newton.
At the end of the nineteenth century a beautiful picture emerged in
the work of mathematician Henri Poincaré, which combined Fermat's
interest in finding rational solutions to algebraic equations with the new
four-dimensional view of elliptic curves: in the four-dimensional picture,
the rational points are perfectly evenly spaced.

For instance, the rational points on the curve $x^3 + y^3 = 1$ are the point
$(1, 0)$, the point $(0, 1)$, and an extra honorary point at infinity. But when
you carry the picture up to the symmetrical torus in four-dimensional
space, these three points become the vertices of an equilateral triangle.

Similarly, the rational points on other elliptic curves will give you
squares, or regular pentagons, or regular hexagons, or pairs of regular
pentagons, or various other things. And sometimes you get infinitely
many rational points on the elliptic curve, and they're smeared out to
appear to fill up a circle or a pair of circles.

The mathematician Louis Mordell, in the first half of this century,
enlarged on Poincaré's vision. In this, he drew inspiration from Fermat's
ideas, including the method that Fermat used in proving FLT for the
case $n = 4$.

As mathematicians continued to study elliptic curves, a dichotomy
emerged. Every elliptic curve either had infinitely many rational points
or else had sixteen or fewer (counting the point at infinity as an honorary
rational point). But no one had a proof of this.

In 1969, mathematician Yves Hellegouarch realized that if you could get an elliptic curve whose rational points, when lifted up to the symmetrical four-dimensional picture of the curve, formed a regular p-sided polygon ("p-gon"), with p a large prime number, you'd be well on your way to getting a counterexample to FLT for exponent p. Turning that around: if you knew that FLT was true for the exponent p, you'd know that you couldn't have p rational points on an elliptic curve arranged in a regular p-gon.

Two other mathematicians working in the 1970s and early 1980s, Vadim Demjanenko and Gerhard Frey, partly independently but with some mutual influence, studied the problem and came to similar conclusions about the connection between elliptic curves and Fermat's Last Theorem. But for these researchers, the link was initially seen as bad news. Elliptic curves were what they wanted to understand; reducing their question to a notoriously difficult problem didn't seem like progress. As Frey wryly put it, "To try to solve a question and to come to Fermat's problem is not encouraging."

But at least there was a link! And in fact Hellegouarch had done some work back in the 1970s, suggesting that the link went both ways—that is, if you could prove the claim about elliptic curves, you might be able to use that information to get a proof (or partial proof) of FLT. Somewhat later, but independently, Frey noticed the same thing: if $A^n + B^n = C^n$ is a counterexample to FLT, then the elliptic curve $y^2 = x(x - A)(x + B)$ has properties that look very fishy to someone conversant with modern number theory. Maybe (Hellegouarch and Frey thought), using known facts about elliptic curves, one could prove that this elliptic curve had self-contradictory properties. Then one would know that no such triple (A, B, C) existed. That is, one could use facts about elliptic curves as a way of tackling FLT.

In the mid-1970s, Barry Mazur found a proof of the result about elliptic curves that Hellegouarch, Demjanenko, and Frey had tried and failed to prove. When Frey heard the news, he was electrified. Maybe Mazur's result, or Mazur's methods, could be applied to prove FLT!

The Road to Wiles' Attic

Frey began to look for more ways to relate FLT to what was known about elliptic curves, as well as what wasn't known but was strongly believed. Like Hellegouarch, Frey studied the properties of an elliptic curve $y^2 = x(x - A)(x + B)$ derived from a putative counterexample to FLT. He worked at the problem for several years, studying it from various angles. Finally, he found the angle that seemed most promising. In 1984 he startled the mathematical community by announcing strong reasons for thinking that such an elliptic curve couldn't be modular.

I haven't told you what it means for an elliptic curve to be "modular," and I'll only explain it here in the vaguest of terms: it means that there's an entirely different way to think about that specific elliptic curve using a different, even weirder geometry, called hyperbolic geometry. By the 1980s most number-theorists believed that all elliptic curves—to be precise, all rational elliptic curves (elliptic curves given by equations involving only rational numbers as coefficients, and having at least one rational solution)—were modular. This proposition had become known as the Shimura–Taniyama–Weil Conjecture, in honor of Yutaka Taniyama, who had first proposed a preliminary version of it, and Goro Shimura and André Weil, who had made the claim more specific and more testable. Subsequent researchers had obtained abundant evidence in favor of the proposition. So for Frey to announce that he'd found a way to construct an elliptic curve that seemed to be non-modular was quite dramatic—even if his construction hinged on FLT being false.

This announcement gave Frey's work an impact that the work of Hellegouarch had lacked. The seemingly non-modular elliptic curves of Hellegouarch and Frey were dubbed "Frey curves," and number-theorists began to study them with the hope of proving that they were as non-modular as they seemed to be (assuming that they existed at all).

Frey's work wasn't a theorem, but more of a sketch, with some key ideas missing. In 1986 Kenneth Ribet, building on work of Jean-Pierre Serre, showed that Frey was right: if there were a counterexample to FLT, then the associated Frey curve would have to be a non-modular elliptic curve.

And *this* convinced many experts, who'd hitherto been agnostic about FLT, that FLT must be was true—because there was so much evidence that every rational elliptic curve was modular.

At this point Andrew Wiles, energized by Ribet's result, decided to try to prove that all rational elliptic curves were modular, or at least all elliptic curves in a broad class that included the Frey curves. This result would give the needed contradiction. That's because Ribet's work had shown that the Frey curve, constructed from a putative counterexample to FLT, *wasn't* modular. If Wiles could show that Frey's curve *was* modular, this contradiction would show that no such curve could exist. That is, no such counterexample A, B, C could exist, and Fermat's Last Theorem would be established!

The Nut, Cracked

So we can say (at last!) that the theory of elliptic curves was the soil in which Wiles wanted to plant the seed of the proof. But the soil wasn't exactly easy to till. Many people badly wanted to know whether all rational elliptic curves were modular, as seemed to be the case, but the experts were convinced that a proof was a long way off. Wiles, in attempting to prove some version of the modularity conjecture (as the linchpin of a proof of FLT), took an odd sort of consolation from the notorious difficulty of the modularity conjecture: at least he wouldn't have to worry that a lot of people were trying the same thing he was working on. The smart money said that it was too soon to try to prove Shimura–Taniyama–Weil.

It turned out that the smart money was, in a way, right: the tools that Wiles needed didn't all exist in the 1980s. But during the period when Wiles was doing his work, other researchers created some of the tools he needed, not realizing the use to which they could be put. Wiles' timing, with hindsight, can be judged to have been nearly perfect: the new tools gave him the leverage he needed just when he needed it. After seven years of hard work, plus an eighth excruciating year of announcement, retraction, collaboration, and revision, Wiles finally proved in 1994 that the (with hindsight, fictitious) Frey curve was modular. In combination with Ribet's work, this proved Fermat's Last Theorem at last.

I want to stress that the twentieth-century proof of Fermat's Last Theorem uses not just algebra and calculus and elliptic curves but all kinds of modern math. So you shouldn't get the idea that the fourth dimension is *the* magic key to the problem; it's one of dozens of magic keys, all of which played crucial roles.

It might seem unjust that such a huge amount of machinery, whose scope I have barely hinted at, should be required for the solution of as simple-sounding a problem as Fermat's Last Theorem. But we all know the principle of leverage that makes a nutcracker work, and it makes a kind of sense that when one is trying to crack as eminently tough a nut as FLT, it might be necessary to apply force at a point far removed from where the nut itself is, or seems to be.

The image of the nutcracker, with its suggestions of strength judiciously applied, is meant to convey a sense of both the effort and the elegance behind Wiles' accomplishment, but the analogy leaves out something important that I tried to convey earlier with a different agricultural metaphor. We shouldn't forget that a nut in the end is just another kind of seed. Perhaps when Fermat's problem is planted in the soil of some still-unknown mathematical country, it will open in the way seeds are designed to open, from the inside out. Then we may have a more accessible answer to this most delightfully accessible and wonderfully difficult of mathematical riddles.

Acknowledgements

Thanks to Joe Buhler, Henry Cohn, John Conway, Noam Elkies, Gerhard Frey, Yves Hellegouarch, Franz Lemmermeyer, Barry Mazur, and Kenneth Ribet for commenting on earlier versions of this article and offering general technical advice. All mistakes are the responsibility of the author.

Bibliography

[Bur81] Stefan Burr. Planting trees. In David A. Klarner, editor, *The Mathematical Gardner*, pages 90–99. Prindle, Weber, and Schmidt, Boston, 1981.

[Hal] http://kresch.com/online_games/asteroids/. A web version of the game Asteroids. Copyright by Mike Hall (1998).

[Key] http://humber.northnet.org/weeks/TorusGames/. Torus-Games copyrighted by Key Curriculum Press (2001).

[Pro] Jim Propp. http://www.math.wisc.edu/~propp/flt4d.html. Some notes and references associated with this article.

[Sch] Rich Schwartz. Lucy and Lily. A four-dimensional "video game". http://www.math.umd.edu/~res/. Copyright by Rich Schwartz (2000).

Games People Don't Play

Peter Winkler

Not all games are to play; some of the most amusing are designed just to think about. Is the game fair? What's the best strategy? The games we describe below were collected from various sources by word of mouth, but thanks to readers of an earlier version we have written sources for some of them. An odd (actually, even) feature of the games in this article is that each has two versions, with entertaining contrasts between the two. There are four pairs of games: the first involving numbers, the second hats, the third cards, and the fourth gladiators. We present all the games first, then their solutions.

The Games

Larger or Smaller (Standard Version)

We begin with a classic game which makes a great example in a class on randomized algorithms. Paula (the perpetrator) takes two slips of paper and writes an integer on each. There are no restrictions on the two numbers except that they must be different. She then conceals one slip in each hand.

Victor (the victim) chooses one of Paula's hands, which Paula then opens, allowing Victor to see the number on that slip. Victor must now guess whether that number is the larger or the smaller of Paula's two numbers; if he guesses right he wins $1, otherwise he loses $1.

Clearly, Victor can achieve equity in this game merely by flipping a coin to decide whether to guess "larger" or "smaller." The question is:

Peter Winkler is Director of Fundamental Mathematics Research at Bell Labs, Lucent Technologies.

not knowing anything about Paula's psychology, is there any way he can do better than break even?

Larger or Smaller (Random Version)

Now let's make things much easier for Victor: instead of being chosen by Paula, the numbers are chosen independently at random from the uniform distribution on [0,1] (two outputs from a standard random number generator will do fine).

To compensate Paula, we allow her to examine the two random numbers and *to decide which one Victor will see.* Again, Victor must decide whether the number he sees is the larger or smaller of the two, with $1 at stake. Can he do better than break even? What are his and Paula's best (i.e., "equilibrium") strategies?

Colored Hats (Simultaneous Version)

Each member of a team of n players is to be fitted with a red or blue hat; each player will be able to see the colors of the hats of his teammates, but not the color of his own hat. No communication will be permitted. At a signal each player will simultaneously guess the color of his own hat; all the players who guess wrong are subsequently executed.

Knowing that the game will be played, the team has a chance to collaborate on a strategy (that is, a set of schemes—not necessarily the same for each player—telling each player which color to guess, based on what he sees). The object of their planning is to *guarantee* as many survivors as possible, assuming worst-case hat distribution.

In other words, we may assume the hat-distributing enemy knows the team's strategy and will do his best to foil it. How many players can be saved?

Colored Hats (Sequential Version)

Again, each of a team of n players will be fitted with a red or blue hat; but this time the players are to be arranged in a line, so that each player can see only the colors of the hats in front of him. Again each player must guess the color of his own hat, and is executed if he is wrong; but this time the guesses are made sequentially, from the back of the line toward the front. Thus, for example, the ith player in line sees the hat-colors of players $1, 2, \ldots, i-1$ and hears the guesses of players $i+1, \ldots, n$

(but he isn't told which of those guesses were correct—the executions take place later).

As before, the team has a chance to collaborate beforehand on a strategy, with the object of guaranteeing as many survivors as possible. How many players can be saved in the worst case?

Next Card Red

Paula shuffles a deck of cards thoroughly, then plays cards face up one at a time, from the top of the deck. At any time Victor can interrupt Paula and bet $1 that the next card will be red. (If he never interrupts, he's automatically betting on the last card.)

What's Victor's best strategy? How much better than even can he do? (Assume there are 26 red and 26 black cards in the deck.)

Next Card Color Betting

Again Paula shuffles a deck thoroughly and plays cards face up one at a time. Victor begins with a bankroll of $1, and can bet any fraction of his current worth, prior to each revelation, on the color of the next card. He gets even odds regardless of the current composition of the deck. Thus, for example, he can decline to bet until the last card, whose color he of course knows, then bet everything and be assured of going home with $2.

Is there any way Victor can *guarantee* to finish with more than $2? If so, what's the maximum amount he can assure himself of winning?

Gladiators, with Confidence-Building

Paula and Victor each manage a team of gladiators. Paula's gladiators have strengths $p_1, p_2, \ldots, p_m$ and Victor's, $v_1, \ldots, v_n$. Gladiators fight one-on-one to the death, and when a gladiator of strength x meets a gladiator of strength y, the former wins with probability $x/(x + y)$ and the latter with probability $y/(x + y)$. Moreover, if the gladiator of strength x wins he gains in confidence and inherits his opponent's strength, so that his own strength improves to $x + y$; similarly, if the other gladiator wins, his strength improves from y to $x + y$.

After each match, Paula puts forward a gladiator (from those on her team who are still alive), and Victor must choose one of his to face Paula's. The winning team is the one which remains with at least one live player.

What's Victor's best strategy? In particular, if Paula begins with her best gladiator, should Victor respond from strength or weakness?

Gladiators, with Constant Strength

Again Paula and Victor must face off in the Coliseum, but this time confidence is not a factor and when a gladiator wins he keeps the same strength he had before.

As before, prior to each match, Paula chooses her entry first. What is Victor's best strategy? Whom should he play if Paula opens with her best man?

Solutions and Comments
Larger or Smaller (Standard Version)

As far as we know, this problem originated with Tom Cover in 1986 and appears as a 1-page "chapter" in his book [Cov87]. Amazingly, there *is* a strategy which guarantees Victor a better than 50% chance to win.

Before playing, Victor selects a probability distribution on the integers which assigns positive probability to each integer. (For example, he plans to flip a coin until a "head" appears. If he sees an even number $2k$ of tails, he will select the integer k; if he sees $2k - 1$ tails, he will select the integer $-k$.)

If Victor is smart he will conceal this distribution from Paula, but as you will see Victor gets his guarantee even if Paula finds out.

After Paula picks her numbers, Victor selects an integer from his probability distribution and adds $1/2$ to it; that becomes his "threshold" t. For example, using the distribution above, if he flips 5 tails before his first head, his random integer will be -3 and his threshold t will be $-2\frac{1}{2}$.

When Paula offers her two hands, Victor flips a *fair* coin to decide which hand to choose, then looks at the number in that hand. If it exceeds t, he guesses that it is the larger of Paula's numbers; if it is smaller than t, he guesses that it is the smaller of Paula's numbers.

So why does this work? Well, suppose that t turns out to be larger than either of Paula's numbers; then Victor will guess "smaller" regardless of which number he gets, and thus will be right with probability exactly $1/2$. If t undercuts both of Paula's numbers, Victor will inevitably guess "larger" and will again be right with probability $1/2$.

But, *with positive probability*, Victor's threshold t will fall *between* Paula's two numbers; and then Victor wins regardless of which hand he picks. This possibility, then, gives Victor the edge which enables him to beat 50%.

Comment. Neither this nor any other strategy enables Victor to guarantee, for some fixed ε, a probability of winning greater than $50\% + \varepsilon$. A smart Paula can choose randomly two consecutive multi-digit integers, and thereby reduce Victor's edge to a smidgeon.

Larger or Smaller (Random Version)

It looks like the ability to choose which number Victor sees is paltry compensation to Paula for not getting to pick the numbers, but in fact *this* version of the game is strictly fair: Paula can prevent Victor from getting any advantage at all.

Her strategy is simple: look at the two random real numbers, then feed Victor the one which is closer to $1/2$.

To see that this reduces Victor to a pure guess, suppose that the number x revealed to him is between 0 and $1/2$. Then the unseen number is uniformly distributed in the set $[0, x] \cup [1 - x, 1]$ and is therefore equally likely to be smaller or greater than x. If $x > 1/2$ then the set is $[0, 1 - x] \cup [x, 1]$ and the argument is the same.

Of course Victor can guarantee probability $1/2$ against any strategy by ignoring his number and flipping a coin, so the game is completely fair.

Comment. This amusing game was brought to my attention only a year ago, at a restaurant in Atlanta. Lots of smart people were stymied, so if you failed to spot this nice strategy of Paula's, you're in good company.

Colored Hats (Simultaneous Version)

It is not immediately obvious that any players can be saved. Often the first strategy considered is "guessing the majority color"; e.g., if $n = 10$, each player guesses the color he sees on 5 or more of his 9 teammates. But this results in 10 executions if the colors are distributed 5-and-5, and the most obvious modifications to this scheme also result in total carnage in the worst case.

However, it is easy to save $\lfloor n/2 \rfloor$ players by the following device. Have the players pair up (say, husband and wife); each husband chooses the color of his wife's hat, and each wife chooses the color she *doesn't* see on her husband's hat. Clearly, if a couple have the same color hats, the husband will survive; if different, the wife will survive.

To see that this is best possible, imagine that the colors are assigned uniformly at random (e.g., by fair coin-flips), instead of by an adversary. Regardless of strategy, the probability that any particular player survives is exactly $1/2$; therefore the expected number of survivors is exactly $n/2$. It follows that the *minimum* number of survivors cannot exceed $\lfloor n/2 \rfloor$.

Colored Hats (Sequential Version)

This version of the hats game was passed to me by Girija Narlikar of Bell Labs, who heard it at a party (the previous version was my own response to Girija's problem, but has no doubt been considered many times before). For the sequential version it is easy to see that $\lfloor n/2 \rfloor$ can be saved; for example, players n, $n-2$, $n-4$ etc. can each guess the color of the player immediately ahead, so that players $n-1$, $n-3$ etc. can echo the most recent guess and save themselves.

It seems like some probabilistic argument such as provided for the simultaneous version should also work here, to show that $\lfloor n/2 \rfloor$ is the most that can be saved. Not so: in fact, all the players except the last can be saved!

The last player (poor fellow) merely calls "red" if he sees an odd number of red hats in front of him, and "blue" otherwise. Player $n-1$ will now know the color of his own hat; for example, if he hears player n guess "red" and sees an *even* number of red hats ahead, he knows his own hat is red.

Similar reasoning applies to each player going up the line. Player i sums the number of red hats he sees and red guesses he hears; if the number is odd he guesses "red," if even he guesses "blue," and he's right (unless someone screwed up).

Of course the last player can never be saved, so $n-1$ is best possible.

Next Card Red

It looks as if Victor can gain a small advantage in this game by waiting for the first moment when the red cards in the remaining deck outnumber the black, then making his bet. Of course, this may never happen and if it doesn't, Victor will lose; does this compensate for the much greater likelihood of obtaining a small edge?

In fact it's a fair game. Not only has Victor no way to earn an advantage, he has no way to lose one either: all strategies are equally effective and equally harmless.

This fact is a consequence of the martingale stopping time theorem, and can also be established without much difficulty by induction (by two's) on the number of cards in the deck. But there is another proof, which I will describe below, and which must surely be in "the book"[1].

Suppose Victor has elected a strategy S, and let us apply S to a slightly modified version of "Next Card Red." In the new version, Victor interrupts Paula as before, but this time he is betting not on the *next* card in the deck, but instead on the *last* card of the deck.

Of course, in any given position the last card has precisely the same probability of being red as the next card. Thus the strategy S has the same expected value in the new game as it did before.

But, of course, the astute reader will already have observed that the new version of "Next Card Red" is a pretty uninteresting game; Victor wins if the last card is red, regardless of his strategy.

There is a discussion of "Next Card Red" in Tom Cover's book [CT91, p. 132–133] on information theory, based on an unpublished result in [Cov74].

Comment. The modified version of "Next Card Red" reminds me of a game which was described—for satiric purposes—in the *Harvard Lampoon*[2] many years ago. Called "The Great Game of Absolution and Redemption," it required that the players move via dice rolls around a Monopoly-like board, until everyone has landed on the square marked "DEATH." So how do you win?

Well, at the beginning of the game you were dealt a card face down from the Predestination Deck. At the conclusion you turn your card face up, and if it says "damned," you lose.

Next Card Color Betting

Finally, we have a really good game for Victor. But can he do better than doubling his money, regardless of how the cards are distributed?

---

[1] As many readers will know, the late, great mathematician Paul Erdős often spoke of a book owned by God in which is written the best proof of each theorem. I imagine Erdős is reading the book now with great enjoyment, but the rest of us will have to wait.

[2] *Harvard Lampoon* Vol. CLVII No. 1, March 30, 1967, pp. 14–15. The issue is dubbed "Games People Play Number" and the particular game in question appears to have been composed by D.C. Kenney and D.C.K. McClelland.

It is useful first to consider which of Victor's strategies are optimal in the sense of "expectation." It is easy to see that as soon as the deck comes down to all cards of one color, Victor should bet everything at every turn for the rest of the game; we will dub any strategy which does this "reasonable." Clearly, every optimal strategy is reasonable.

Surprisingly, the converse is also true: no matter what Victor's strategy is, as long as he comes to his senses when the deck becomes monotone, his expectation is the same! To see this, consider first the following *pure* strategy: Victor imagines some fixed specific distribution of red and black in the deck, and bets *everything he has* on that distribution *at every turn*.

Of course, Victor will nearly always go broke with this strategy, but if he wins he can buy the earth—his take-home is then $2^{52} \times \$1$, around 50 quadrillion dollars. Since there are $\binom{52}{26}$ ways the colors can be distributed in the deck, Victor's mathematical expected return is $\frac{\$2^{52}}{\binom{52}{26}} = \$9.0813$.

Of course, this strategy is not realistic but it is "reasonable" by our definition, and most importantly, *every reasonable strategy is a combination of pure strategies of this type*. To see this, imagine that Victor had $\binom{52}{26}$ people working for him, each playing a different one of the pure strategies.

We claim that every reasonable strategy of Victor's amounts to distributing his original \$1 stake among these assistants, in some way. If at any time his collective assistants bet $\$x$ on "red" and $\$y$ on black, that amounts to Victor himself betting $\$(x - y)$ on "red" (when $x > y$) or $\$(y - x)$ on black (when $y > x$).

Each reasonable strategy yields a distribution, as follows. Say Victor wants to bet \$.08 that the first card is red; this means that the assistants who are guessing "red" first get a total of \$.54 while the others get only \$.46. If, on winning, Victor plans next to bet \$.04 on black, he allots \$.04 more of the \$.54 total to the "red-black" assistants than to the "red-red" assistants. Proceeding in this manner, eventually each individual assistant has his assigned stake.

Now, any ("convex") combination of strategies with the same expectation shares that expectation, hence every reasonable strategy for Victor has the same expected return of \$9.08 (yielding an expected profit of \$8.08). In particular all reasonable strategies are optimal.

But one of these strategies *guarantees* \$9.08; namely, the one in which the \$1 stake is divided equally among the assistants. Since we can never guarantee more than the expected value, this is the best possible guarantee.

```
0                                                                    101 202 404 808
1                                                                    202 303 404 404
2                                                                190 253 303 303 202
3                                                         96 134 178 222 253 253 202 101
4                                                        132 167 200 222 222 190
5                                                    101 129 158 184 200 200 178
6                                                    126 150 171 184 184 167 134
7                                                123 144 161 171 171 158 132  96
8                                            101 120 138 153 161 161 150 129
9                                         83 100 117 133 146 153 153 144 126 101
10                                         99 115 129 140 146 146 138 123
11                                      98 112 125 135 140 140 133 120
12                                   97 110 121 130 135 135 129 117 101
13                                96 108 118 126 130 130 125 115 100
14                             95 106 115 122 126 126 121 112  99  83
15                          94 104 113 119 122 122 118 110  98
16                 73  83  93 102 110 116 119 119 115 108  97
17                 83  92 101 108 113 116 116 113 106  96
18                 91  99 106 111 113 113 110 104  95
19                 98 104 109 111 111 108 102  94
20  74  82  90  97 103 107 109 109 106 101  93
21  82  89  96 101 105 107 107 104  99  92  83
22  89  95 100 103 105 105 103  98  91  83  73
23  94  99 102 103 103 101  97
24  98 101 102 102 100  96  90
25 100 101 101  99  95  89  82
26 100 100  98  94  89  82  74

   26  25  24  23  22  21  20  19  18  17  16  15  14  13  12  11  10   9   8   7   6   5   4   3   2   1   0
```

Figure 1. Optimal strategy in the discrete 100 cent Next Card Color game.

Comment. This strategy is actually quite easy to implement (assuming as we do that U.S. currency is infinitely divisible). If there are b black cards and r red cards remaining in the deck, where $b \geq r$, Victor bets a fraction $(b - r)/(b + r)$ of his current worth on black; if $r > b$, he bets $(r - b)/(r + b)$ of his worth on red.

If the original \$1 stake is *not* fungible, but is composed of 100 indivisible cents, things become more complicated and it turns out that Victor does about a dollar worse. A dynamic program (written by Ioana Dumitriu of M.I.T.) shows that optimal play by Victor and Paula results in Victor ending with \$8.08; Figure 1 shows the size of Victor's bankroll at each stage of a well-played game. For example, if the game reaches a point when there are 12 black and 10 red cards remaining, Victor should have \$1.08. By comparing the entries above and to the right we see that he should bet either \$.11 (in which case Paula will let him win) or \$.12 (in which case he will lose) that the next card is black.

Note that Victor tends to bet slightly more conservatively in the "100 cents" game than in the continuous version. If instead he chooses to bet always the nearest number of cents to the fraction $(b - r)/(b + r)$ of his current worth, Paula will knock him down to \$0 before half the deck is gone!

I heard this problem from Russ Lyons, of Indiana University, who heard it from Yuval Peres, who heard it from Sergiu Hart; Sergiu doesn't remember where he heard it but suspects that Martin Gardner may have written about it decades ago!

Gladiators, with Confidence-Building

As in "Next Card Red," all strategies for Victor are equally good.

To see this, imagine that strength is money. Paula begins with $P = p_1 \ldots p_m$ dollars and Victor with $V = v_1 \ldots v_n$ dollars. When a gladiator of strength x beats a gladiator of strength y, the former's team gains \$$y$ while the latter's loses \$$y$; the total amount of money always remains the same. Eventually, either Paula will finish with \$$P$ + \$$V$ and Victor with zero, or the other way 'round.

The key observation is that every match is a fair game. If Victor puts up a gladiator of strength x against one of strength y, then his expected financial gain is

$$\frac{x}{x + y} \cdot \$y \; + \; \frac{y}{x + y} \cdot (-\$x) \; = \; \$0 \; .$$

Thus the whole tournament is a fair game, and it follows that Victor's expected worth at the conclusion is the same as his starting stake, \$$P$. Thus

$$q(\$P + \$V) + (1 - q)(\$0) = \$P$$

where q is the probability that Victor wins. Thus $q = P/(P + V)$, independent of anyone's strategy in the tournament.

Comment. Here's another, more combinatorial, proof, pointed out by one of my favorite collaborators, Graham Brightwell of the London School of Economics. Using approximation by rationals and clearing of denominators, we may assume that all the strengths are integers. Each gladiator is assigned x balls if his initial strength is x, and all the balls are put into a uniformly random vertical order. When two gladiators battle the one with the higher topmost ball wins (this happens with the required $x/(x+y)$ probability) and the loser's balls accrue to the winner.

The surviving gladiator's new set of balls is again uniformly distributed in the vertical order, just as if he had started with the full set; hence the outcome of each match is independent of previous events, as required. But regardless of strategy, Victor will win if and only if the top ball in the whole order is one of his; this happens with probability $P/(P+V)$.

Gladiators, with Constant Strength

Obviously, the change in rules makes strategy considerations in this game completely different from the previous one—or does it? No, again the strategy makes no difference!

For this game we take away each gladiator's money (and balls), and turn him into a lightbulb.

The mathematician's ideal lightbulb has the following property: its burnout time is completely memoryless. That means that knowing how long the bulb has been burning tells us absolutely nothing about how long it will continue to burn.

The unique probability distribution with this property is the exponential; if the expected (average) lifetime of the bulb is x, then the probability that it is still burning at time t is $e^{-t/x}$.

Given two bulbs of expected lifetimes x and y, respectively, the probability that the first outlasts the second is—you guessed it—$x/(x+y)$. We imagine that the matching of two gladiators corresponds to turning on the corresponding lightbulbs until one (the loser) burns out, then turning off the winner until its next match; since the distribution is memoryless, the winner's strength in its next match is unchanged.

During the tournament Paula and Victor each have exactly one lightbulb lit at any given time; the winner is the one whose total lighting time (of all the bulbs/gladiators on her/his team) is the larger. Since this has nothing to do with the order in which the bulbs are lit, the probability that Victor wins is independent of strategy. (Note: that probability is a more complex function of the gladiator strengths than in the previous game).

Comment. The constant-strength game appears in [KLN84]. I have a theory that the other game came about in the following way: someone enjoyed the problem and remembered the answer (all strategies equally good) but not the conditions. When he or she tried to reconstruct the rules of the game, it was natural to introduce the inherited-strength condition in order to make a martingale.

Afterword

It seems only right that I conclude with a game whose solution is left to the reader. Described by Todd Ebert in his 1998 U.C. Santa Barbara Ph.D. thesis,[3] it can be thought of as another version of the (simultaneous) colored hats game. In this version the colors really are chosen by independent fair coin-flips; each of 15 players will get to see the colors of all the other players' hats, and has the *option* of guessing the color of his own hat. There is to be no communication between the players; in particular, no player can tell what color a teammate has guessed or even whether he has guessed at all.

The players conspire beforehand and must come up with a strategy which maximizes the probability that *every* guess is correct, subject to the condition that it must guarantee that at least one player guesses.

As usual, there is an elegant solution and proof of optimality. Hint: the players can attain a 50% chance of all-correct by appointing one player to guess and the rest to pass. It's hard to believe, on first sight, that they can do any better; but in fact they can beat 90%!

Acknowledgement

I am obviously indebted to the (mostly) unknown inventors of the games described above, not to mention the many other wonderful mathematical brainteasers that have come my way over the years. If you are the originator, or know the originator, of any of them, I will be most grateful for a communication.

Bibliography

[Cov74] T. Cover. Universal gambling schemes and the complexity measures of Kolmogorov and Chaitin. Statistics Department 12, Stanford University, October 1974.

---

[3]Readers can find the problem in T. Ebert and H. Vollmer, "On the Autoreducibility of Random Sequences," http://www-info4.informatik.uni-wuerzburg.de/person/mitarbeiter/vollmer/publications.html.

[Cov87] T. Cover. Pick the largest number. In T. Cover and B. Gopinath, editors, *Open Problems in Communication and Computation*, page 152. Springer Verlag, 1987.

[CT91] T. Cover and J. Thomas. *Elements of Information Theory*. Wiley, 1991.

[KLN84] K.S. Kaminsky, E.M. Luks, and P.I. Nelson. Strategy, nontransitive dominance and the exponential distribution. *Australian Journal of Statistics*, 26(2):111–118, 1984.

Mathematical Chats Between Two Physicists

Aviezri S. Fraenkel

To Martin Gardner—The Master of recreational mathematics

The Luncheon Chat

Joyce is a physicist doing statistical mechanics, and Gill a nuclear physicist specializing in particle interactions. While relaxing with their cups of coffee after a tasty enjoyable light lunch at the TEX (TasteEnjoyrelaχ)—the *Sciences Club* of the University—they began to chat about some common aspects of their specialties.

Gill: The interaction between elements such as particles, nucleons, spins, etc., that are "close" to one another is common to our two disciplines. I wonder whether a lesson can be learned by viewing these phenomena in a unified manner.

Joyce: Hmm...a nice idea. I think that to do this we need some abstract model that reflects the basic common properties of these interactions, and that is amenable to mathematical analysis, such as working with two elements 1 and 0, that form a field called by those pompous mathematicians the *Galois field* of two elements, GF(2).

G: Yes, GF(2) has the advantage that $1 = -1$, so the rule $1 + 1 = 0$ in this field is the same as the annihilation rule of particles and spins: $1 - 1 = 0$. We have of course $0 + 1 = 1 + 0 = 1$ and $0 + 0 = 0$,

Aviezri S. Fraenkel is a scholar and computicianeer—computer scientist, mathematician and engineer, who worked on the design of one of the earliest digital computers and has fathered the Responsa Retrieval Project.

as well as $1 + 1 = 0$. These addition rules are also known as *Nim sum* or *Xor— exclusive or*. Furthermore, to model interactions that are not necessarily neighboring vertices on a grid, it seems best to have a directed graph $G = (V, E)$—that mathematicians, always tending to succinctness, call *digraph* for short—on whose vertices V the "particles" 0 and 1 are initially distributed. Selecting a particle on a vertex u, it is complemented as well as all its neighbors along edges directed away from u.

J: What you describe is a system called *cellular automata* by those inflated logicians, mathematicians, and computer scientists, a manifestation of which is the *Merlin Magic Square* game manufactured by Parker Brothers (but Arthur-Merlin games are something else again). Quite a bit is known about such solitaire games. Anyway, a huge literature has been accumulating on cellular automata. A small example, intersecting with solitaire games, is [Gol91], [Pel87], [Sto89], [Sut88], [Sut89], [Sut90], [Sut95]. Incidentally, related but different solitaires are *chip firing games*, see, e.g., [BL92], [Lóp97], [Big99].

What seems more attractive and new is to transform these solitaire games into two-player games, where the player first achieving 0s on all the non-leaf vertices wins and the opponent loses. If there is no last move, the outcome is a draw. Moreover, this version will appeal to many of my colleagues who have turned their attention to biology, such as protein folding, where the main aim is to tinker with nature, in order to achieve some doubtful benefits such as designing specialized medicines and genetic engineering (alias tinkering)...For want of a better name, we might call them *Cellata* games, since it reminds me both of the Italian cuisine that I just enjoyed, and of cellular automata.

G (*taking a paper napkin and beginning to draw on it*): I like your idea, and I share your belief that it appears to be new and interesting. In most of the solitaire games you have mentioned, *any* order of the moves produces the same result. To promote your suggestion of tinkering, I think it's then best to permit the players to select only an *occupied* vertex, i.e., a vertex occupied by a 1. So a move in the game consists of selecting an occupied vertex and *firing* it, i.e., complementing it together with all its directed neighbors. The player making the last move wins. If there is no last move, the outcome is a draw...the order of the moves is then definitely important, unlike in those solitaires. ...Here now is a suggested game on two components with an initial 0, 1-distribution, where 1s are indicated by ⋆s (Figure 1) and vertices occupied by 0s remain unlabeled. As a gentleman, I'm used to "Ladies First" etiquette, so I graciously offer you to move first.

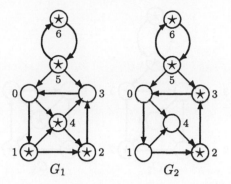

Figure 1. A two-player game $G_1 + G_2$ on cellular automata. A move consists of selecting a vertex v marked with a $\star$ and "firing" it. Once fired, the $\star$ is removed, and $\star$s are placed on every vertex v points to. If two $\star$s appear at a vertex, both are annihilated. Two players play by taking turns firing a vertex. The first player unable to move loses, and the opponent wins. If there is no last move, the outcome is a draw. The result of firing vertex 4 in G_1 is shown in G_1 of Figure 2.

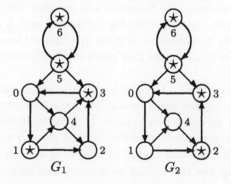

Figure 2. Game $G1 + G2$ from Figure 1 after one move.

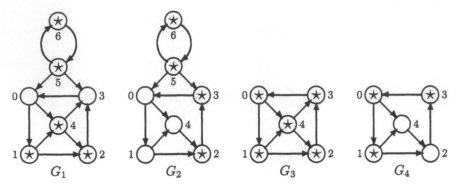

Figure 3.   Adding two more components, G_3 and G_4.

J (*pulling a PalmCrash from her handbag and hammering away furiously on its buttons*): You propose to play a *sum* of games, i.e., a move consists of selecting a component and firing an occupied vertex on it. The player making the last move in the entire digraph wins, and her opponent loses...it seems to me that your gentlemanly gesture is all but gallant. It is indeed patronizing, since whatever I'll do from this position, you can win. I'll therefore add to your two components two simplified versions, namely deleting vertices 5 and 6 on the two components, with ⋆s as indicated (Figure 3). Under these circumstances I accept your offer to make the first move in the sum consisting of all the 4 components.

G (*blushing*): Well...I really hadn't expected you to find out so soon...I see that on the game consisting of the four components you can win by making an appropriate move.... Since it seems that both of us understand the win/lose positions of this game, I suggest to play the same game with the small change of adjoining a ⋆ on vertex 0 of G_1.

J (*consulting her PalmCrash once more and then rising*): All right, the initial position is now a draw. Since we seem to have mastered also the draw positions, it's time to head back to our offices and do some serious physics...such as deciding the computational complexity of Cellata games.

A Conversation in Joyce's Office

The next day, Professor Gill Andrin strolled over to Professor Joyce Prato's office.

Gill: Good morning Joyce, I was wondering how you found me out so quickly yesterday when I offered you to play first on Figure 1.

Joyce: Hi Gill, I'm already used to your tricks. When I saw that you proposed to play on two components of a game that obviously has cycles, I assumed that you had computed the generalized Sprague–Grundy function γ for the game [Smi66], [Con76, Ch. 11], [FY86]; otherwise you would hardly be able to beat a sharp opponent and be so smug about it. (Walking over to the whiteboard.) I suspected that γ is *additive* (also called *linear*) on the digraph $\mathbf{G} = (\mathbf{V}, \mathbf{E})$, induced by the given ground-graph $G = (V, E)$, where $\mathbf{V}$ is the collection of all subsets of vertices from $V = (z_1, \ldots, z_n)$. That is, $\gamma(\mathbf{u}) \oplus \gamma(\mathbf{v}) = \gamma(\mathbf{u} \oplus \mathbf{v})$ whenever either $\gamma(\mathbf{u}) < \infty$ or $\gamma(\mathbf{v}) < \infty$. The $\oplus$ denotes Nim sum, and every $\mathbf{w} \in \mathbf{V}$ is an n-dimensional binary vector with 1s precisely in locations i where z_i is an occupied vertex in G. I proved linearity with the aid of my PalmCrash. This enabled me to compute γ very easily.

G: Congratulations. But how could you possibly prove linearity with the aid of a computer?

J: I took lots of examples, and it always confirmed linearity. There was no counterexample at all.

G: Hmm...Is this a standard method of proof in statistical mechanics?

J: Well, I don't need the formal proofs of those highbrow mathematicians. I perceive truth when I meet it.

G: It appears that you have been a little hard on mathematicians, especially yesterday. Many phenomena are counterintuitive. I concur with the mathematicians that proofs of claims are necessary, though the precise notion of "proof" might be debatable. Of course one might formulate a *conjecture*, and base further results on it.

To come back to our Cellata game, $\gamma(u)$, when finite, is the smallest nonnegative integer not appearing among the *options* (direct followers) of vertex u...instead of using n-dimensional vectors to denote vertices of $\mathbf{V}$, it will now be more convenient to denote them by $n_1 \ldots n_k$, where $z_{n_1}, \ldots, z_{n_k}$ are the occupied vertices of V. Thus you presumably noticed that on G_1, $\gamma(4) = 0$, since it has the as yet unlabeled option 23, that has the option Φ, the configuration with no $\star$s, for which obviously $\gamma(\Phi) = 0$. Similarly, $\gamma(02) = \gamma(13) = 0$. Using linearity, we then get

$$\mathbf{V}_0 = \{\Phi, 4, 02, 13, 024, 134, 0123, 01234\},$$

where $\mathbf{V}_i$ is the subset of $\mathbf{V}$ on which γ assumes the value i ($i < \infty$). In fact, γ is a homomorphism from $\mathbf{V}^f$ (the linear subspace of the vector space $\mathbf{V}$ on which γ is finite) onto $\mathrm{GF}(2)^t$ for some nonnegative integer

t with kernel $\mathbf{V}_0$ and quotient space $\mathbf{V}^f/V_0 = \{\mathbf{V}_i : 0 \leq i < 2^t\}$, and $\dim(\mathbf{V}^f) = t + \dim(\mathbf{V}_0)$. We have $\mathbf{V}^\infty = \mathbf{V} \setminus \mathbf{V}^f$, where $\mathbf{V}^\infty$ is the subset on which $\gamma = \infty$. For G_1, $\gamma(23) = 1$, since its only options are $\{\Phi, 02\} \subseteq \mathbf{V}_0$. Also $\gamma(56) = 2$. We thus get the cosets

$$\mathbf{V}_1 = 23 \oplus \mathbf{V}_0 = \{23, 234, 03, 12, 034, 124, 01, 014\},$$

$\mathbf{V}_2 = 56 \oplus \mathbf{V}_0$, $\mathbf{V}_3 = 0356 \oplus \mathbf{V}_0$, $\dim \mathbf{V}_0 = 3$, $\dim \mathbf{V}^f = 5$, $t = 2$.

For G_2 we get

$$\mathbf{V}_0 = \{\Phi, 1, 02, 34, 012, 134, 0234, 01234\},$$

$\mathbf{V}_1 = 23 \oplus \mathbf{V}_0$, $\mathbf{V}_2 = 56 \oplus \mathbf{V}_0$, $\mathbf{V}_3 = 0356 \oplus \mathbf{V}_0$, $\dim \mathbf{V}_0 = 3$, $\dim \mathbf{V}^f = 5$, $t = 2$.

It follows that the γ-value on G_1 is $\gamma(56) \oplus (124) = 2 \oplus 1 = 3$, and also on G_2 we have a γ-value of 3. Their Nim sum is thus 0, which means that whoever moves from this position loses. Is this how you figured things out?

J: Precisely. For G_3 and G_4 that I adjoined to the game, we have $\mathbf{V}_0$, $\mathbf{V}_1$ as for G_1 and G_2 respectively, but $\dim \mathbf{V}_0 = 3$, $\dim \mathbf{V}^f = 4$, $t = 1$. Therefore on G_3, $\gamma(01234) = 0$ and on G_4, $\gamma(013) = \gamma((23) \oplus (012)) = 1$. Thus firing vertex 0 on G_4, results in 34, with $\gamma(34) = 0$. This is a winning move, since γ now vanishes on the entire digraph.

G: Yes. By adjoining a $\star$ at vertex 0 in G_1, we get $\gamma(012456) = \infty$, so the sum of the four components has also γ-value infinity, and the outcome is now a draw, as you said. I better leave now, as I got to teach my Graduate Mesoscopic Physics course.

J: Enjoy—bye.

The Truncated Chat in the Faculty Room

Joyce and Gill met again next day in the Faculty room where dough-nuts, cookies, coffee, and tea were served in anticipation of an important gathering.

Joyce (*moving to the whiteboard*): I thought it would be interesting to change the rules, a particular case of which would be to fire the selected vertex u and complement precisely any *two* of its options in the groundgraph if $d_{\text{out}}(u) \geq 2$; and complement all the options of u if $d_{\text{out}}(u) \leq 2$. (I'm now using the terminology "firing" in a new sense: complementing the selected vertex and some subset of its options.) I conjecture that additivity holds also for this game. The digraph $G(s)^2$

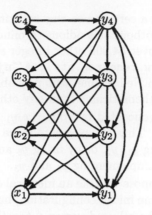

Figure 4. Playing on a parametrized digraph.

on which I would like to play this game depends on a parameter $s \in \mathbb{Z}^+$. It has vertex set $\{x_1, \ldots, x_s, y_1, \ldots, y_s\}$, and edges:

$$
\begin{aligned}
F(x_i) &= y_i \quad \text{for } i = 1, \ldots, s, \\
F(y_k) &= \{y_i : 1 \le i < k\} \cup \{x_j : 1 \le j \le s \text{ and } j \ne k\} \quad \text{for } k = 1, \ldots, s.
\end{aligned}
$$

As an example, I'm drawing $G(4) = G(4)^2$ on the board (Figure 4). Suppose we play on $G(7)$, and place 1s precisely on the 8 vertices $x_7, y_1, \ldots, y_7$. Can you figure out the nature of this position?

Gill: (*fingering the knobs of the WallComp next to the whiteboard*): Before doing that, why didn't you consider the G^1 version, i.e., firing an occupied vertex (in your new sense of "firing"), and complementing precisely *one* of its options?

J: Well, this would be a pure particle physics game without much appeal to statistical mechanics, and it was *you* who had suggested to consider a unified approach. Besides, this special case was solved in [Fra74], [FY76], [FY82], where a polynomial strategy was formulated. The misère version was analyzed in [Fer84].

G: I expected you to say this, but it gave me time to think about the question you asked me...I concur with your conjecture about additivity. It seems that though the groundgraph $G(s)$ has no leaf, the game-graph $G(s)$ has no γ-value ∞. It also appears that any collection of x_i is in $\mathbf{V}_0$. The value of the y_i seem to be more tricky...I think that $\gamma(y_i) =$ the ith *odious* number, where the odious numbers are those positive integers whose binary representations have an odd number of 1-bits. Incidentally, odious numbers arise in the analysis of other games, such as Grundy's game, Kayles, Mock Turtles, Turnips. See [BCG82].

They arose earlier in a certain two-way splitting of the nonnegative integers [LM59] (but without this odious terminology...). More information about this and over 54,000 other integer sequences is available online from `http://www.research.att.com/~njas/sequences/` Thanks to the mathematician Neil Sloane, who probably contributed more to a larger number of mathematicians than any other mathematician!

For the position concocted by you, $\gamma(x_7 y_1 \ldots y_7) = 0 \oplus 1 \oplus 2 \oplus 4 \oplus 7 \oplus 8 \oplus 11 \oplus 13 = 14$. Thus the player to move can win: either by firing y_7 and complementing y_1, y_2, or by firing y_6 and complementing y_1, y_3, or by firing y_5 and complementing y_2 and y_3.

J: Very nice...suppose we take an identical clone of $G(s)^2$, and begin with precisely the same initial configuration, but change the rule for the clone: complement the selected vertex u together with any *three* of its options if $d_{\text{out}}(u) \geq 3$; and all the options of u if $d_{\text{out}}(u) \leq 3$. We better call this new clone $G(s)^3$, to distinguish it from $G(s)^2$. Can the first player win also here?

G (*moving to within reach of both the whiteboard and the WallComp*): Let's see...on the clone, all collections of an even number of x_i are in $\mathbf{V}_0$; and $\mathbf{V}^f$ consists precisely of all collections of an *even* number of $1s$...we seem to have $\gamma(x_j y_j) = $ smallest nonnegative integer not the Nim sum of at most three $\gamma(x_i y_i)$ for $i < j$. Thus $\gamma(x_7 y_1 \ldots y_7) = 0 \oplus 1 \oplus 2 \oplus 4 \oplus 8 \oplus 15 \oplus 16 \oplus 32 = 48$. So firing y_7 and complementing y_6 and any two of the x_i ($i < 7$) is a winning move. ...Incidentally, the sequence $\{1, 2, 4, 8, 15, 16, 32, 51, \ldots\}$ appears also in Neil's Encyclopædia, and has been used in [BCG82] for a special case of the game "Turning Turtles."

J: How about playing the sum of $G(s)^2$ and $G(s)^3$ with the same given initial position on both clones?

G: That's easy. The value of the sum is simply the Nim sum of their γ-values which is $14 \oplus 48$. To win we have to move in $G(s)^3$ to a position with γ-value 14. There is a unique winning move of changing the γ-value 32 to 30. This is affected by firing y_7 and complementing y_6, y_5 and y_1...I hear in the corridor the President talking with the Cabinet Minister of Science approaching...we better adjourn before we'll have to explain to the minister that we are playing a game.

J (*moving to the WallComp*): Not before we briefly summarize where we stand...We still should address the question of the computational complexity of Cellata games...and yes, I concede that it would be nice to prove additivity formally for the family of all Cellata games...In these games, is every draw position necessarily such that *every* move from it leads to another draw? This is the case for all the games we considered, but it would be nice to provide a case where this doesn't hold...We played *impartial* games. How about playing a sort of *partizan* game on, say, $G(s)^3$ and $G(s)^2$ simultaneously, i.e., one player follows the $G(s)^3$ rules and her opponent the $G(s)^2$ rules?...I think I can see some interesting applications in fields other than physics. Incidentally, the case of $G(s)^4$, where a vertex on G_s is fired and any *four* of its options are complemented, seems to give rise to the sequence $1, 2, 4, 8, 16, 31, 32, 64, 103, \ldots$. It is the sequence $\gamma(x_j y_j)$ defined as the smallest nonnegative integer not the Nim sum of at most four earlier terms. This sequence was not in the Encyclopædia of integers, so I just sent a message, via the WallComp, to your latest mathematics hero Neil Sloane, together with the fact that it appears in Table 3, Chapter 14 of [BCG82]. Note that the strategy of our Cellata games on just Figure 4 alone subsumes and unifies that of a battery of games there...I just noted that Sloane has added the new sequence into his Encyclopædia.

If it wouldn't be for our own University President who seeks to elicit more money from this narrow-minded minister, I'd proudly tell the latter that we are playing a game, followed by a quote from the founder of our "Sciences Club":

"...A third purpose of this book is to have fun. Indeed, pleasure has probably been the main goal all along. But I hesitate to admit it, because computer scientists want to maintain their image as hard-working individuals who deserve high salaries. Sooner or later society will realize that certain kinds of hard work are in fact admirable even though they are more fun than just about anything else." ([Knu93b, p. iii], see also [Knu77].)

Bibliography

[BCG82] E. R. Berlekamp, J. H. Conway, and R. K. Guy. *Winning Ways for your Mathematical Plays* (volumes I and II). Academic Press, London, 1982. Translated into German: *Gewinnen, Strategien für Mathematische Spiele* by G. Seiffert, Foreword by K. Jacobs, M. Reményi and Seiffert, Friedr. Vieweg & Sohn, Braunschweig (four volumes), 1985.

[Big99] N. L. Biggs. Chip-firing and the critical group of a graph. *Journal of Algebraic Combinatorics*, 9(1):25–45, 1999.

[BL92] A. Björner and L. Lovász. Chip-firing games on directed graphs. *Journal of Algebraic Combinatorics*, 1(4):305–328, 1992.

[Con76] John H. Conway. *On Numbers and Games*. Academic Press, London/New York, 1976. Translated into German: *Über Zahlen und Spiele* by Brigitte Kunisch, Friedr. Vieweg & Sohn, Braunschweig, 1983.

[Fer84] Thomas S. Ferguson. Misère annihilation games. *Journal of Combinatorial Theory. Series A*, 37:205–230, 1984.

[Fra74] Aviezri S. Fraenkel. Combinatorial games with an annihilation rule. In J. P. LaSalle, editor, *The Influence of Computing on Mathematical Research and Education (Proc. Symp. Appl. Math., Vol. 20, Univ. Montana, 1973)*, pages 87–91. American Mathematical Society, Providence, RI, 1974.

[FY76] A. S. Fraenkel and Y. Yesha. Theory of annihilation games. *Bulletin of the American Mathematical Society*, 82(5):775–777, 1976.

[FY82] A. S. Fraenkel and Y. Yesha. Theory of annihilation games—I. *Journal of Combinatorial Theory. Series B*, 33(1):60–86, 1982.

[FY86] A. S. Fraenkel and Y. Yesha. The generalized Sprague–Grundy function and its invariance under certain mappings. *Journal of Combinatorial Theory. Series A*, 43(2):165–177, 1986.

[Gol91] Eric Goles. Sand piles, combinatorial games and cellular automata. *Mathematics and its Applications*, 64:101–121, 1991.

[Knu77] D. E. Knuth. Are toy problems useful? *Popular Computing*, 5:3–10, 1977.

[Knu93b] D. E. Knuth. *The Stanford GraphBase: A Platform for Combinatorial Computing*. ACM Press, New York, 1993.

[LM59] J. Lambek and L. Moser. On some two way classifications of integers. *Canadian Mathematical Bulletin*, 2:85–89, 1959.

[Lóp97] C. M. López. Chip firing and the Tutte polynomial. *Annals of Combinatorics*, 1(3):253–259, 1997.

[Pel87] D. H. Pelletier. Merlin's magic square. *American Mathematical Monthly*, 94(2):143–150, 1987.

[Smi66] C. A. B. Smith. Graphs and composite games. *Journal of Combinatorial Theory*, 1:51–81, 1966. Reprinted in slightly modified form in: *A Seminar on Graph Theory* (F. Harary, ed.), Holt, Rinehart and Winston, New York, NY, 1967.

[Sto89]   D. L. Stock. Merlin's magic square revisited. *Amererican Mathematical Monthly*, 96(7):608–610, 1989.

[Sut88]   K. Sutner. On σ-automata. *Complex Systems*, 2(1):1–28, 1988.

[Sut89]   K. Sutner. Linear cellular automata and the Garden-of-Eden. *Mathematical Intelligencer*, 11(2):49–53, 1989.

[Sut90]   K. Sutner. The σ-game and cellular automata. *American Mathematical Monthly*, 97(1):24–34, 1990.

[Sut95]   K. Sutner. On the computational complexity of finite cellular automata. *Journal of Computer and System Science*, 50(1):87–97, 1995.

How Flies Fly: The Curvature and Torsion of Space Curves

Rudy Rucker

Thanks for everything, Martin. You're the best ever.

It's interesting to watch flies buzz around. They trace out curves in space that are marvelously three-dimensional. Birds fly along space curves too, but their airy swoops are not nearly so bent and twisted as are the paths of flies.

Is there a mathematical language for talking about the shapes of curves in space? Sure there is. Math is the science of form, and mathematicians are always studying nature for new forms to talk about.

Historically, space curves were first discussed by the mathematician Alexis-Claude Clairaut in a paper called "Recherche sur les Courbes a Double Courbure," published in 1731 when Clairaut was eighteen [Kli72, p. 557]. Clairaut is said to have been an attractive, engaging man; he was a popular figure in eighteenth-century Paris society.

In speaking of "double curvature," Clairaut meant that a path through three-dimensional space can warp itself in two independent ways; he thought of a curve in terms of its shadow projections onto, say, the floor and a wall. In discussing the bending of the planar, "shadow" curves, Clairaut drew on then recent work by the incomparable Isaac Newton.

Newton's mathematical curvature measures a curve's tendency to bend away from being a straight line. The more the curve bends, the greater is the absolute value of its curvature. From the viewpoint of a point moving along the curve, the curvature is said to be positive when

Rudy Rucker has published 21 popular science and science fiction books.

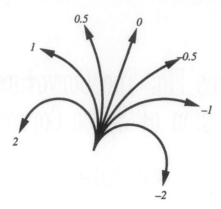

Figure 1.   Curvature along circular arcs in the plane.

the curve bends to the left, and negative when the curve bends to the right. The size of the curvature is determined by the principle that a circle of radius R is defined to have a curvature of $1/R$. The smaller the radius, the greater the curvature. Figure 1 shows some examples of circular arcs, with their curvatures indicated.

We often represent a curve in the plane by an equation involving x and y coordinates. Most calculus students remember a brief, nasty encounter with Newton's formula for the curvature of a curve; the formula uses fractional powers and the first and second derivatives of y with respect to x. Fortunately, there is no necessity for us to trundle out this cruel, ancient idol. Instead we think of curvature as a primitive notion and express the curve in a more natural way.

The idea is that instead of talking about positions relative to an arbitrary x-axis and y-axis, we think of a curve as being a bent number-line by itself. The curve is marked off in units of "arclength," where arclength is the distance measured along the curve, just as if the curve were a piece of rope that you could stretch out next to a ruler. We'll use the variable s to stand for arclength and the infinitesimal ds to stand for a very small bit of arclength.

If we think of a curve in x and y coordinates, we can define ds as the square root of dx squared plus dy squared, and we can then use integration to add up the ds quantities to get a value for s. But in this essay, we'll instead think of s and ds as primitive quantities. If we think of the arclength s as primitive, the most natural way to describe a plane curve is by an equation that gives the curvature directly as a function of arclength, an equation of the form $\kappa = f(s)$, where κ is the commonly used symbol for curvature. Figure 2 shows two famous plane curves

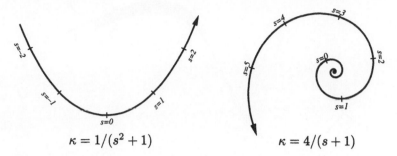

$$\kappa = 1/(s^2 + 1) \qquad\qquad \kappa = 4/(s + 1)$$

Figure 2. The catenary and the logarithmic spiral expressed by natural equations, with curvature κ a function of arclength s. Arclength is marked as units along the curves.

which happen to have simple expressions for curvature as a function of arclength. The catenary curve is the shape assumed by a chain (or bridge cable) suspended from two points, while the logarithmic spiral is a form very popular among our friends the mollusks.

Note that for the spiral, the center is where s approaches -1. And if you jump over the anomalous central point and push down into larger negative values of s, you produce a mirror-image of the spiral.

It would be nice to also think of space curves in a natural, coordinate-free way—surely this is the way a fly buzzing around in the center of an empty room must think. Profound mathematical insights come hard, and it was a hundred and twenty years after Clairaut before the correct way to represent a space curve by intrinsic natural equations was finally discovered—by the French mathematicians Joseph Alfred Serret and Frederic-Jean Frenet.

The idea is that at each point of a space curve one can define two numerical quantities called curvature and torsion. The curvature of a space curve is essentially the same as the curvature of a plane curve: it measures how rapidly the curve is bending to one side. The torsion measures a curve's tendency to twist out of a plane. But what exactly is meant by "bend to one side," and "twist out of a plane"? Which plane?

The idea is that at each point P of a space curve you can define three mutually perpendicular unit-length vectors: the tangent T, the normal N, and the binormal B. T shows the direction the curve is moving in, N lies along the direction which the curve is currently bending in, and B is a vector perpendicular to T and N. (In terms of the vector cross product, $T \times N$ is B, $N \times B$ is T, and $B \times T$ is N.) For space curves we ordinarily work only with positive values of curvature, and have N

Figure 3.   The moving trihedron of a space curve: T the tangent, N the normal, and B the binormal.

point in the direction in which the curve is actually bending. (In certain of the analytical curves we'll look at later we relax this condition and allow negative curvature of space curves.)

Taken together, T, N, and B make up the so-called "moving trihedron of a space curve." In Figure 3 we show part of a space curve with two instances of the moving trihedron. So that it's easier to see the three-dimensionality of the image, we draw the curve as a ribbon like a twisted ladder. The curve runs along one edge of the ladder, and the rungs of the ladder correspond to the directions of successive normals to the curve.

To understand exactly how the normal is defined, it helps to think of the notion of the "osculating" (kissing) plane. At each point of a space curve there is some plane that best fits the curve at that point. The tangent vector T lies in this plane, and the direction perpendicular to T in this plane holds the normal N. The binormal is a vector perpendicular to the osculating plane.

With the idea of the moving trihedron in mind, we can now say that the curvature measures the rate at which the tangent turns, and the torsion measures the rate at which the binormal turns.

Note that T, N, and B are always selected so as to form a right-handed coordinate system. This means that if you hold out the thumb, index finger and middle finger of your right hand, these directions correspond to the tangent, the normal, and the binormal.

Just as the circle is the plane curve characterized by having constant curvature, the helix is the space curve characterized by having constant curvature and constant torsion. Figure 5 shows how the signs of the curvature and torsion affect the shapes of plane and space curves.

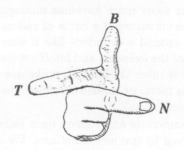

Figure 4.   A right-hand as a trihedron.

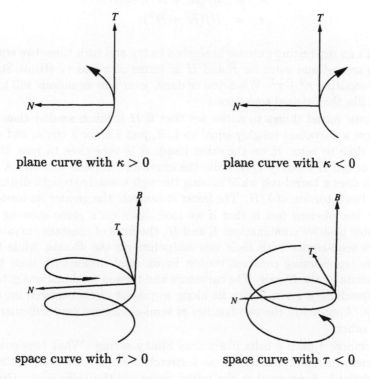

plane curve with $\kappa > 0$ plane curve with $\kappa < 0$

space curve with $\tau > 0$ space curve with $\tau < 0$

Figure 5.   How the signs of the curvature and torsion affect the motion of a curve.

Now let's look for some space formulae analogous to the plane formula stating that the curvature of a circle of radius R is $1/R$. Think of a helix as wrapping around a cylinder—like a vine growing up a post. Let R be the radius of the cylinder, and let H represent the turn-height: the vertical distance it takes the helix to make one complete turn (and to make the formulae nicer, we measure turn-height in units 2π as large as the units we measure R in.)

The sizes of the curvature and torsion on a helix with radius R and turn-height H are given by two nice equations. We write "τ" for torsion and, as before, "κ" for curvature:

$$\kappa = R/(R^2 + H^2), \text{ and}$$
$$\tau = H/(R^2 + H^2).$$

It's an interesting exercise in algebra to try and turn these two equations around and solve for R and H in terms of κ and τ. (Hint: Start by computing $\kappa^2 + \tau^2$. When you're done, your new equations will look a lot like the original equations.)

Some initial things to notice are that if H is much smaller than R, you get a curvature roughly equal to $1/R$, just like for a circle, and a τ very close to zero. If, on the other hand, R is very close to zero, then the torsion is roughly $1/H$ while the curvature is close to zero. A fly which does a barrel-roll while moving through a nearly straight distance of H has a torsion of $1/H$. The faster it can roll, the greater its torsion.

A less obvious fact is that if we look down on a plane showing all possible positive combinations R and H, the lines of constant curvature lie on semi-circles with their two endpoints on the R-axis; while the points representing constant torsion lie on semi-circles with their two endpoints on the H-axis. The curvature and torsion combinations gotten by stretching a given Slinky lie along a quarter circle centered on the origin. Apparently the two families of semi-circles are perpendicular to each other.

Suppose I have a helix like a steel Slinky spring. What happens to the curvature and the torsion as I stretch a single turn of it without untwisting? Suppose that the initial radius of the helix is A. Given the physical fact that the length of one twist of the Slinky keeps the same length, you can show that as you stretch it, $R^2 + H^2$ will stay constant at a value of A^2, which corresponds to a circle of radius A around the origin of the R–H plane. As you stretch a Slinky loop with the particular starting radius of 2, its R and H values will move along the dashed line shown in Figure 6. Figure 7 shows what a few of the

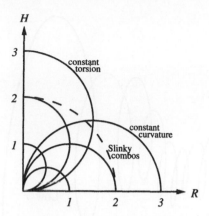

Figure 6. Lines of constant curvature and torsion for combinations of R (helix radius) and H (helix turn height).

intermediate positions will look like. Curvature is being traded off for torsion.

Here's a little algebra problem: Given the formulae for κ and τ in terms of R and H, and given that $R^2 + H^2 = A^2$, what can you say about the sum $\kappa^2 + \tau^2$? The answer tells you more about the nature of a Slinky's trade-off between curvature and torsion.

One fact that seems odd at first is that the curvature and torsion of a helix are dependent on the size of the helix. If you make both R and H five times as big, you make the torsion and curvature $1/10$ as big. If you make R and H N times as big, you make the curvature and torsion $1/(2*N)$ as big.

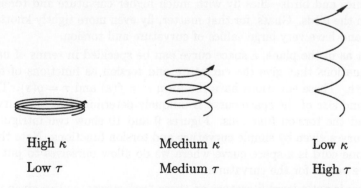

| High κ | Medium κ | Low κ |
| Low τ | Medium τ | High τ |

Figure 7. Stretching a Slinky turns curvature into torsion.

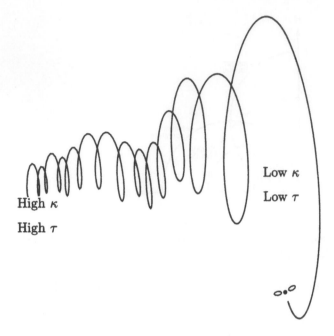

Figure 8.   Changing curvature and torsion.

But this makes sense if you think of a fly that switches from a small helix to a big helix; the fly is indeed changing the way that it's flying, so it makes sense that the κ and the τ should change.

This observation suggests a simple way to express the difference between flies and birds—flies fly with much higher curvature and torsion than do the birds. Gnats, for that matter, fly even more tightly knotted paths, and have very large values of curvature and torsion.

Just as in the plane, a space curve can be specified in terms of natural equations that give the curvature and torsion as functions of the arclength. These equations have the form $\kappa = f(s)$ and $\tau = g(s)$. The shape and size of the space curve is uniquely determined by the curvature and the torsion functions. Figures 9 and 10 show two intriguing space curves given by simple curvature and torsion functions. Note that the phone cord is a space curve where we do allow ourselves to put in negative values for the curvature.

There is not a large literature on these "$\kappa\tau$" curves, so I've given my own names to these two: the rocker, and the phone cord.

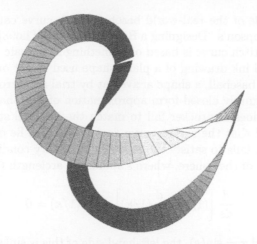

Figure 9.   The rocker, with natural equations $\kappa = 1$ and $\tau = \sin(\text{arclength})$.

Figure 10.   The phone cord, with natural equations $\kappa = 10\sin(\text{arclength})$ and $\tau = 3$.

At one time I thought that the rocker was a correct way to represent the seam on a tennis ball or the stitching on a baseball, but an email from the mathematician John Horton Conway convinced me I was wrong. Conway makes the anthropological conjecture that every time a mathematician discovers a curve that he or she thinks might be the true baseball curve, the curve is a different one!

An analysis of the real-world baseball stitch curve can be found in Richard Thompson's "Designing a Baseball Cover" [Tho98]. It turns out the baseball stitch curve is based on something so prosaic as a patented 1860s pen and ink drawing of a plane shape used to cut out the leather for a half of a baseball, a shape arrived at by trial and error. Thompson finds a fairly gnarly closed-form approximation of this shape.

Not only does my rocker fail to match the baseball stitch curve, it can be proved that the rocker curve does not in fact lie on the surface of a sphere. It fails to satisfy the following necessary condition for lying on the surface of the sphere, where s stands for arclength (see [Won72]).

$$\frac{d}{ds}\left[(1/\tau)\cdot\frac{d}{ds}(1/\kappa)\right]+\tau\cdot(1/\kappa)=0$$

(For $\kappa = 1$ and $\tau = \sin(s)$, the left-hand side of this is $\sin(s)$, which isn't identically 0.)

Numerical estimates indicate that the arclength of the rocker has exactly twice the length of a circle of the same radius. This suggests an easy way to make a rocker. Cut out two identical annuli (thick circles) from some fairly stiff paper (manila file folders are good), cut radial slits in the annuli, tape two of the slit-edges together, bend the annuli in two different ways (one like a clockwise helix and one like a counterclockwise helix) and tape the other two slit-edges together, forming a continuous band of double length. Because an annulus cannot bend along its osculating plane, the curvature of the shape is fixed along the arclength. Because half the band is like a clockwise helix and half is like a counterclockwise helix, when the shape relaxes, the torsion presumably varies with the arclength like a sine wave function that goes between plus one and minus one. The torsion seems to be zero at the two places where the slits are taped together. Note that I have not proved that my empirical paper rocker is the same as my mathematical rocker, this is simply my conjecture.

- To make the rocker, make a (larger) copy of Figure 11 on stiff paper.

- Cut along all lines.

- Tape edge A to edge B^* with the letters on the same side.

- Bend the two rings in the opposite sense.

- Tape edge A^* to edge B with the letters on the same side.

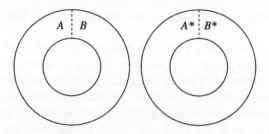

Figure 11. Make your own rocker.

How were the images in Figures 9 and 10 generated? They use an algorithm based on the 1851 formulae of Serret and Frenet (see, for instance, [Str61]). Let's state the formulae in "differential" form. The question the formulae address is this: when we do a small displacement ds along a space curve, what is the displacement dT, dN, and dB of the vectors in the moving trihedron?

$$dT = (\qquad\quad \kappa N \qquad\quad) \cdot ds$$
$$dN = (-\kappa T \qquad\quad +\tau B) \cdot ds$$
$$dB = (\qquad\quad -\tau N \qquad\quad) \cdot ds$$

The first and third equations correspond, respectively, to the definitions of curvature and torsion. The second equation describes the "back-reaction" of the T and B motions on N.

A good aid to remembering the Frenet formulae is to note that if we think of the ds multipliers on the righthand sides of the three equations as linear combinations of T, N, and B, then the coefficients in these combinations make a three-by-three antisymmetric matrix, that is, a matrix in which the ij-entry is the negative of the ij-entry.

Since we are lucky enough to live in three-dimensional space, it is possible for us to experiment with our bodies and to perceive directly why the Serret–Frenet formulae are true. To experience the equations, you should, if possible, stick out your right hand's thumb, index finger, and middle finger as shown in Figure 4. Now start trying to "fly" your trihedron around according to these rules: (1) The index finger always points in the direction your hand is moving. (2) You are allowed to turn the index finger towards or away from direction of the middle finger by a motion corresponding to rotating around the axis of your thumb. (3) You are allowed to turn the thumb towards or away from the middle finger by a motion corresponding to rotating around the axis of your forefinger.

To get clear on what's meant by motion (2), grab your thumb with your left hand and make as if you were trying to unscrew it from your hand. This is a kind of "yawing" motion, and it corresponds to the first of the three Serret–Frenet formulae: the change in the tangent is equal to the curvature times normal. Motion (3) corresponds to grabbing your index finger with your left hand and trying to unscrew that finger. This is a kind of "rolling" motion, and it corresponds to the third of the Serret–Frenet formulae: the change in the binormal is the negative of the torsion times the normal.

In thinking of flying along a space curve you should explicitly resist thinking about boats and airplanes which have a built-in visual trihedron which generally does not correspond to the moving trihedron of the space curve. If you do want to think about a machine, imagine a rocket which never slows down and never speeds up, which can turn left or right— relative to you the passenger—and which can roll. Or better yet, think about being a cybernetic house-fly.

An exciting thing about the Serret–Frenet formulae is that they lend themselves quite directly to creating a numerical computer simulation to create $\kappa\tau$ space curves with arbitrary curvature and torsion. To write the code in readable form, we can "overload" the arithmetic operators to do the expected things to our vector objects. A scalar times a vector changes the length of the vector, while a vector times a vector invokes the vector cross product. In addition we add a vector function called `Normalize` such that if a vector `A` invokes the method by calling `A.Normalize()`, then `A` becomes a unit vector. Here is the heart of an algorithm for updating the position P of a point on an arbitrary $\kappa\tau$ curve.

```
P = P + ds * T;
s = s + ds;
T = T + (kappa(s) * ds) * N;
B = B + (-tau(s) * ds) * N;
T.Normalize();
B.Normalize();
N = (B * T);
```

As far as I know, very little mathematical work has been done with $\kappa\tau$ curves because in the past nobody could visualize them. I first implemented the algorithm as a Mathematica notebook for the Macintosh and for Windows machines, and then I wrote a stand-alone Windows

Figure 12. A $\kappa\tau$ curve with curvature varying as a random walk.

program called Kaptau. You can download either of the Mathematica notebooks or the stand-alone Windows program from my web-site [Ruc].

So how do I think flies fly? I think that they generally move along at a constant speed like a space curve parameterized by its arclength, and that they manage to loiter here and speed away from there by varying their curvature and torsion between low and high values. As mathematicians like to say (even when they're wrong): "It's obvious!"

Bibliography

[Kli72] Morris Kline. *Mathematical Thought from Ancient to Modern Times.* Oxford University Press, 1972.

[Ruc] Rudy Rucker. http://www.mathcs.sjsu.edu/faculty/rucker.

[Str61] Dirk Struik. *Lectures on Classical Differential Geometry.* Addison-Wesley, Reading, Massachusetts, 1961. Also available in a paperback reprint from Dover Books.

[Tho98] Richard B. Thompson. Designing a baseball cover. *The College Mathematics Journal*, 29(1), January 1998. A related article is at http://www.mathsoft.com/asolve/baseball/baseball.html.

[Won72] Yung-Chow Wong. On an explicit characterization of spherical curves. *Proceedings of the American Mathematical Society*, 34:239–242, 1972.

Some Tricks and Paradoxes

Raymond Smullyan

Some Mischievous Tricks

I was planning to play a particularly mischievous trick on my dear friend, Martin Gardner, at the banquet of the 1998 Gardner gathering, but unfortunately Martin couldn't attend due to a temporary indisposition of his lovely wife Charlotte. And so, I played almost the same trick (the same except for the ending) on someone else at the banquet. First, I will tell you the trick I *did* play, and then the trick I *would have* played, had Martin attended.

What I did was this: I showed the audience two envelopes and explained that one of them contained a dollar bill and the other one was empty. To make it possible to deduce which envelope contained the bill, I had a sentence written on the outside of each envelope. I then said that if anybody could correctly deduce which envelope contained the dollar bill, then he or she could have the bill. However, for the privilege of being allowed to do this, I would charge 25 cents. Is anyone game?

Here are the two sentences:

(1)	(2)
The sentences on the two envelopes are both false.	The dollar bill is in the other envelope.

Raymond Smullyan is a mathematical logician, magician, musician, essayist, and author of numerous books that introduce deep mathematical topics through recreational logic puzzles.

Can the reader figure out which envelope contains the dollar? Well, one gentleman took me up, and so he then owed me a quarter. I asked which envelope contained the bill. He said it was Envelope #1. I asked his reason. He explained: "If Sentence 1 were true, then both sentences would be false (as Sentence 1 says), which is clearly absurd. Therefore, Sentence 1 cannot be true; it must be false. Since it is false, then what it says is not the case—it is not the case that both sentences are false, so at least one must be true, and since it is not Sentence 1, it must be Sentence 2. Since Sentence 2 is true, then the bill must be in Envelope 1, as the sentence says."

"Very good," I said (giving myself an imaginary mysterious wink). "Now, please open the envelopes." He opened Envelope 1, and found it empty. Then, sure enough, a dollar bill was found in Envelope 2!

"Well, well! I said. "The reasoning sure sounded good, yet clearly something must be wrong somewhere. What?"

Problem 1 *What was wrong with the reasoning?*[1]

At this point the gentleman owed me 25 cents. "I'll tell you what," I said. "Today I'm in a generous mood, and so I'll give you back your quarter if you agree to answer a *yes/no* question truthfully. Fair enough?" He agreed. I then framed a question such that the only way he could answer truthfully is by his paying me a million dollars!

Problem 2 *What question would work?*

"Well, well, well!" I said to him, after he owed a million dollars. "I really feel sorry for you, and quite guilty at having swindled you out of a million dollars, so here is what I'm going to do: I'll give you a chance to win your million dollars back again, but for this privilege I charge a nickel extra. Agreed?" Of course, he agreed, and so he then owed me a million dollars, plus a nickel.

I then said: "In fact, I'm in *such* a generous mood, that I'll give you back the entire million dollars, and even the nickel, providing that you answer a *yes/no* question—now, don't get alarmed!—a *yes/no* question, but this time you don't have to answer truthfully; you may lie or be truthful; your answer can be either true or false, as you choose. Under these circumstances, there is certainly no way I can con you, is there?" It appeared not, and so he agreed. Yet, I framed the question such that he had no option but to pay me a billion dollars![2]

---

[1] Solutions to problems are given at the end of the article.

[2] This is one of my best puzzles and belongs to the field of *coercive logic*, as explained in my book, *The Riddle of Scheherazade* [Smu98].

Problem 3 *What question would work?*

They say, "All's well that ends well," and so after he owed me a billion dollars, I said: "I'll tell you what. I think I'll give you back the billion dollars as a gift and claim a tax deduction." This ended my presentation.

Well, a few days after the conference, I phoned Martin to find out how Charlotte was doing, and told Martin what I would have done, had he been present. I would have done the above to Martin up to the point where he would have owned me a billion dollars, but instead of offering to give it back to him as a gift and claiming a tax deduction, I was planning to say: "I'll tell you what, Martin. I'll trade you the entire billion dollars for one kiss from Charlotte!"

Pretty mischievous, eh? Perhaps, even more mischievous were my escapades during my student days. Upon meeting any attractive girl, I would say, "I'll bet you I can kiss you without touching you." Now, this works particularly well if the girl is highly intelligent and analytical and asks me for precise definitions of *kissing* and *touching*. After I had given them, she would usually say that it was obviously impossible. We would then make the bet, after explaining that there was no money involved; it was simply a bet of honor. I would then ask her to close her eyes, and when she did so, I would kiss her and say: "I lose!"

Pretty mischievous, huh? Actually, my most mischievous trick to date led to a particularly happy conclusion. About 40 years ago, I met a charming lady musician in New York City. On our first date, I asked her if she would agree to the following: I would make a statement, and if the statement were true, she was to give me her autograph. She said, "I don't see why not." I then added, "But if the statement were false, you must promise not to give me your autograph!" She agreed. I then made a statement such that to keep her word, she had to give me, not her autograph, but a kiss!

Problem 4 *What statement would work?*

Well, instead of collecting the kisses, she agreed to play for double or nothing; and so she soon owed me 2 kisses, then 4, then 8, then 16, and so we kept doubling and doubling and things escalated and escalated, the end result of which was that we were soon married![3]

---

[3]Last year, Blanche and I celebrated our 40th anniversary.

Some Paradoxes

Do any of you know the L. A. A. computing company? Do you know what "L. A. A." stands for? It stands for: "Lacking an acronym."

I didn't invent this. I heard it and soon after realized that it is actually no paradox; it is simply *false!* However, I thought of a way of modifying it so that it becomes a paradox. Consider the L. A. C. A. company where "L. A. C. A." stands for "lacking a correct acronym." Is this acronym correct or not? It is certainly an acronym, but is it correct? Assuming that the company has no other acronym, then if it is correct, what it says is really the case, which means that the company has no correct acronym, contrary to the assumption that the acronym is correct. Thus, it is contradictory to assume that the acronym is correct. On the other hand, suppose that the acronym is not correct. Then, since the company has no other acronym, it lacks a correct acronym, which is just what the acronym says, and hence, the acronym is correct after all, which is again a contradiction, and thus the acronym is neither correct nor incorrect, but paradoxical.

I am fond of the **businessman's paradox** due to Lisa Collier: The president of a certain company offered a reward of $100 to any employee who could offer a suggestion which would save the company money. One employee suggested: "Eliminate the reward."

For years I have been searching for what might be termed **a meta-paradox**, i.e., something that is paradoxical if and only if it isn't! Well, I think I have found one:

> Either this sentence is false, or (this sentence is paradoxical if and only if it isn't).

I leave the proof to the reader.

Newcomb's Paradox Without a Predictor.   To those who don't know Newcomb's paradox, I present it in the following diabolical manner: Suppose you are in front of a chest containing two drawers and either each drawer contains a hundred dollars, or each one contains a thousand dollars. You have the choice of either taking the money in both drawers, or just the money in the bottom drawer. Which would you choose? Which choice would yield you the most money? Of course, everyone would opt for both drawers. I then ask whether there is any further information I could give that would make you change your mind and believe that you would find more money if you open just the bottom drawer than if you open both drawers? This sounds impossible, doesn't it? Surely, many of you would be willing to bet that *no* added information could make you change your mind on *that!* But, ah! What I didn't tell you is that there exists and omniscient being—either a human, a computer, or a god— who is a **perfect** predictor and at any time knows the entire future of the universe. This perfect predictor knew in advance how you would choose, and was in complete control of how much money was to go into the drawers. If the being predicted that you would choose both drawers, then $100 was to be put into each drawer, but if it was determined that you would choose just the bottom drawer, then $1000 was to be put in each drawer.[4] Would this added information change your mind?

Well, people are divided into two camps on this: Some say, "Of course, this would change my mind! According to this added information, if I choose both drawers, I will get only $200, whereas if I choose just the bottom drawer, then I will get $1000, and so I should certainly choose just the bottom drawer." But others say, "This would not change my mind in the least! The plain simple fact is that the money is already there, and there is twice as much money in both drawers as there is in just the bottom drawer, hence if I choose both drawers, I will get twice as much than if I choose just the bottom drawer."

Well, dear reader, how do you feel about this? I once discussed this with Martin Gardner and told him that I was definitely in the first camp. Martin (perhaps playing Devil's advocate) then told me that I shouldn't dismiss the other argument so easily! To make the point more dramatic, he said, "Suppose that the back of the chest was made of glass and that friends of yours could see how much money was in each drawer and were hoping that you would choose both drawers. What then?" I don't recall what I answered, but what I *should* have answered, and am now answering is, this: "Assuming that my friends believed the predictor, then they would *know* how I would choose. If they saw $100 in each drawer, then they would know that I would choose both drawers, and if they saw $1000 in each drawer, then the only way they could hope that I would choose both drawers is by hoping that the predictor was

---

[4] Actually, I changed the dollar amounts from the original version to make it more tempting to choose both drawers.

wrong, because if I chose both drawers, the predictor *would* have been wrong!"[5]

We have now seen two arguments—one that you should choose both drawers and one that you should choose just the bottom drawer. Some writers have proposed that this paradox proves that a perfect predictor cannot exist! For interesting discussions of this, see Martin Gardner's discussion in [Gar86a, pp. 155–175].

I totally disagree with this! Indeed, my purpose in writing this is to show that the essential idea behind the paradox can be reformulated in such a manner that the predictor is left out entirely! And so, here is my version:

Again, you are to choose the contents of either both drawers or just the bottom drawer of a chest of two drawers. Now, consider the following proposition:

Proposition 1 *Either you will choose both drawers and there will be $100 in each drawer, or you will choose just the bottom drawer and there will be $1000 in each drawer.*

Please note that the above proposition makes no reference to any predictor! Newcomb's version with the predictor *implies* the above proposition, the proposition itself is far more general. The predictor, of course, gives plausibility to the above proposition, but the proposition can be stated without reference to any predictor.

Now, to those of you who believe in choosing both drawers, I wish to address the following key question: If you believed the above proposition, would you still choose both drawers? I can't see how you could! Of course, I can understand that you would simply have no reason to believe that proposition, but once you believe it, I cannot understand how you would than choose both drawers! As we are at it, is the proposition even consistent? Those of you who take the position that it is logically impossible for there to be a perfect predictor (as the paradox suggests) should also believe that the above proposition is inconsistent. And so, is the proposition consistent or not? Well, modeling the two arguments, I will first prove that the proposition is inconsistent, and then I will prove that it is consistent! That's pretty paradoxical, isn't it? And I will forget the predictor entirely!

---

[5]Incidentally, in some versions of Newcomb's paradox, the predictor is not always correct, but only correct with high probability. This is a totally different problem that I don't want to get into. To know what to do in this version, I would have to know the exact probability that the predictor was right. And so, I will continue to assume that the predictor is always right.

Proof that the proposition is inconsistent. Regardless of whether or not there is $100 in each drawer or $1000 in each drawer, there is twice as much money in both drawers as there is in just the bottom drawer. Therefore, you will get more money by choosing both drawers than by choosing just the bottom drawer.

On the other hand, the above proposition clearly implies that if you choose both drawers, you will get $200, whereas if you choose just the bottom drawer, you will get $1000. Since $1000 is obviously more than $200, you will get more by choosing just the bottom drawer. We have now reached a clear contradiction, hence the proposition is inconsistent.

Proof that the proposition is consistent. To prove that a proposition is consistent, it suffices to show that there is a possibility in which it can be true (since an inconsistent proposition can never be true). Well, it is certainly possible that you choose both drawers and find $100 in each, which would validate the proposition. (Alternately, it is equally possible that you choose just the bottom drawer and that there be $1000 in each drawer.) Thus, the proposition is consistent.

Now obviously, a proposition cannot be both consistent and inconsistent, hence one of the above arguments must be fallacious. Which one?

Problem 5 *Which argument is wrong, and precisely where is the fallacy?*

Solutions

Answer 1 *(Problem 1 on page 342)* To begin with, some of the most experienced logicians have fallen for this fiendish trick of mine! People naturally assume that any given sentence must be either true or false, and this is simply not so! If Sentence 1 were true, we would obviously have a logical contradiction, so it can't be true. It also isn't false, for if it were, the dollar would have to be in Envelope 1, which it isn't. Thus, Sentence 1 is neither true nor false, and the whole error of the argument was the tacit assumption that it was. (Incidentally, Sentence 2 is either true or false—it is well-defined and happens to be false.)

The whole purpose of this game was to dramatically illustrate Tarski's discovery that in a language like English, truth within the language is not definable within the language.

Answer 2 *(Problem 2 on page 342)* The question I used was, "Will you either answer *no* to this question, or pay me a million dollars?" He couldn't correctly answer *no*; he had to answer *yes,* and the only way that this answer could be correct is by paying me a million dollars.

Answer 3 *(Problem 3 on page 343)* The idea is to frame a question such that unless he paid me a billion dollars, neither his *yes* or *no* answer could be either true or false, but would have to be paradoxical! Remember, I never said that he could answer *yes* or *no* at random; his answer had to be either *true* or *false!* And so, the question must be such that the only way he could avoid answering paradoxically is by paying me a billion dollars. A question that works is: "Is *yes* the correct answer to this question if and only if you pay me a billion dollars?" If he didn't pay me a billion dollars, the question would then be equivalent to: "Is *yes* *not* the correct answer to his question?" and neither *yes* nor *no* could be either a *true* or a *false* answer to that question.

Answer 4 *(Problem 4 on page 343)* What I said was: "You will give me neither your autograph nor a kiss." If the statement were true, she would have to give me her autograph, as agreed, but her doing so would falsify the statement (it would be false that she gave me neither her autograph, nor a kiss), hence that statement cannot be true; it must be false. Since it is false that she will give me *neither*, she must give me *either*—she must give me either her autograph or a kiss. But she agreed not to give me her autograph for a false statement, hence she owed me a kiss! (Sneaky, eh?)

Answer 5 *(Problem 5 on page 347)* I would say that the first argument is wrong. It is obvious that the proposition is consistent, so where is the fallacy in the first argument? The fallacy is this: It is indeed true that however you choose, there will be twice as much money in both drawers as in just the bottom drawer, but from this it simply does not follow that you will get twice as much money by choosing both drawers than by choosing just the bottom drawer. You obviously won't, because the amount in the drawers is tied up with how you choose!

Bibliography

[Gar86a] Martin Gardner. *Knotted Doughnuts and Other Mathematical Entertainments.* W.H. Freeman, New York, 1986.

[Smu98] Raymond Smullyan. *The Riddle of Scheherazade and Other Amazing Puzzles, Ancient & Modern.* Harcourt Brace, San Diego, 1998.

Part VII
Mathematical Treats

How Recreational Mathematics Can Save the World

Keith Devlin

Recreational mathematics is generally presented—and is always pursued —for recreational purposes. Just as the name suggests. But is it really possible to draw a line between mathematics pursued for fun and mathematics pursued with applications in mind? Could a piece of recreational mathematics lead to important applications? To take the most extreme case, could recreational math save the world? Could Martin Gardner, the undisputed king of recreational mathematicians, lay claim to have played a role in saving humanity?

Certainly, useful applications of recreational mathematics abound. The role played in modern cryptography by factoring large numbers into primes is the example that perhaps comes most obviously to mind—an example in which Martin's *Scientific American* column "Mathematical Games" played a major catalytic role.

Whether public key cryptography could be said to have the potential for saving the world, however, is another matter. For that accolade, you'd have to find an application along the lines of helping to eliminate widespread famine in a world facing a massive population growth.

Fanciful? Perhaps. But let's just see how we can do. I'll start with one of the best known pieces of recreational mathematics there is, and an old favorite of Martin: the Fibonacci sequence.

Keith Devlin is the author of 24 books, most recently *The Math Gene: How Mathematical Thinking Evolved and Why Numbers Are Like Gossip.*

Once upon a Time There Were Two Rabbits

The story begins in the early 13th century, when the great Italian mathematician Fibonacci (Leonardo of Pisa) posed the following simple problem. A man puts a pair of baby rabbits into an enclosed garden. Assuming that each pair of rabbits in the garden bears a new pair every month, which from the second month on itself becomes productive, how many pairs of rabbits will there be in the garden after one year? Like most mathematics problems, you are supposed to ignore such realistic happenings as death, escape, impotence, or whatever. Given those assumptions, it is not hard to see that the number of pairs of rabbits in the garden in each month is given by the numbers in the sequence 1, 1, 2, 3, 5, 8, 13, etc. This sequence of numbers is called the Fibonacci sequence. The general rule that produces it is that each number after the second one is equal to the sum of the two previous numbers. (So $1 + 1 = 2$, $1 + 2 = 3$, $2 + 3 = 5$, etc.) This corresponds to the fact that each month, the new rabbit births consists of one pair to each of the newly adult pairs plus one pair for each of the earlier adult pairs. Once you have the sequence, you can simply read off that after one year there will be 377 pairs.

As Martin and others have observed on many occasions, the Fibonacci numbers have some curious mathematical properties. Perhaps the most amazing is that, as you proceed along the sequence, the ratios of the successive terms gets closer and closer to the famous *golden ratio* number 1.61803..., the "perfect width-to-height" ratio beloved by the ancient Greeks and incorporated into much of their architecture. (For example, the front of the Parthenon building in Athens has the proportions of the golden ratio.)

To digress briefly, I have to say that I never found the golden rectangle the particularly pleasing shape many textbooks claim it to be, and I know I am not alone in that view. A few years ago, I performed an experiment: I presented a class of college students with a sheet of paper on which I had drawn a number of rectangles of various aspect ratios, among them a golden rectangle, and asked them to choose the one they found the most attractive. Hardly anyone picked the golden rectangle. Instead, they plumped for the one I thought they would: the one having the shape of a modern television screen.

My reason for suspecting this outcome was that, since birth, those students had been regularly exposed to this particular shape. Indeed, it had framed much of their experience of the world. Not surprisingly, therefore, their minds found it highly pleasing.

By the same reasoning, I have no difficulty in believing that ancient Greeks really did find the golden ratio the most pleasing aspect ratio of a rectangle. Because the ancient Greeks deliberately incorporated the golden rectangle in much of their architecture, the children of that television-free society were bombarded with golden rectangles. I suspect it's a matter not of some inbuilt propensity of the human brain, but of regular exposure to the environment. When the proposed wide-screen televisions eventually take over from the existing design, I suspect the "most pleasing rectangle" will have the aspect ratio of the new screens. Since that particular ratio happens to be very close to the golden rectangle, that will make all those textbook remarks true once again.

From Rabbits to Flowers to Feeding the World

To return to my main theme, most popular expositions of mathematics observe that the Fibonacci numbers arise frequently in nature. For example, if you count the number of petals in various flowers you will find that the answer is often a Fibonacci number (much more frequently than you would get by chance). For instance, an iris has 3 petals, a primrose 5, a delphinium 8, ragwort 13, an aster 21, a daisy 34, and Michaelmas daisies 55 or 89 petals—all Fibonacci numbers.

For another example from the botanical world, if you look at a sunflower you will see a beautiful pattern of two spirals, one running clockwise, the other counterclockwise. Count those spirals and you will find that there are 21 running clockwise and 34 counterclockwise—both Fibonacci numbers. Similarly, pine cones have 5 clockwise spirals and 8 counterclockwise spirals; and the pineapple has 8 clockwise spirals and 13 going counterclockwise.

These are not numerological coincidences. Rather they are consequences of the way plants grow. Since the nineteenth century, mathematicians speculated that it was the connection with the golden ratio that gave rise to the appearance of the Fibonacci numbers in the botanical world. In 1993, two French mathematicians, Stephane Douady and Yves Couder, finally proved that this was the case. They showed that the most efficient way to pack petals or seeds next to one another in a growing plant is to stagger them according to the golden ratio. Since the numbers of petals or seeds has to be a whole number, this can only be done approximately, of course. Because the Fibonacci numbers are the whole numbers in a growth pattern that come closest to being "staggered" by the golden ratio, those are the numbers the ever-efficient Mother Nature uses.

Another appearance of the Fibonacci numbers in the natural world arises in the way the leaves are located on the stems of trees and plants. If you take a look, you will see that, in many cases, as you progress up along a stem, the leaves are located on a spiral path that winds around the stem. The spiral pattern is sufficiently regular that it leads to a numerical parameter characteristic for the species, called its divergence. Start at one leaf and let p be the number of complete turns of the spiral before you find a second leaf directly above the first, and let q be the number of leaves you encounter going from that first one to the last in the process (excluding the first one itself). The quotient p/q is the *divergence* of the plant.

If you calculate the divergence for different species of plants, you find that both the numerator and the denominator tend to be Fibonacci numbers. In particular, $1/2$, $1/3$, $2/5$, $3/8$, $5/13$, and $8/21$ are all common divergence ratios. For instance, common grasses have a divergence of $1/2$, sedges have $1/3$, many fruit trees (including the apple) have a divergence of $2/5$, plantains have $3/8$, and leeks come in at $5/13$.

Once again, this is not a coincidence. The leaves on a plant stem should be situated so that each has a maximum opportunity of receiving sunlight, without being obscured by other leaves. The optimal way to do this is for the angle between successive leaves to be the golden ratio.

And now we have our (potential) link between recreational mathematics and feeding a growing population. There is no doubt that it was the observations of recreational mathematicians of the ubiquitous appearance of the Fibonacci numbers of plants and flowers that led mathematicians and botanists to figure out why this was the case. Thus, recreational mathematics played a significant role in developing our understanding of plant growth.

Of course, not all research into plant growth has been motivated by mathematical considerations. But some has, and in the long run it's the total understanding achieved that counts. And without doubt, increased understanding of plant growth processes have led plant scientists to develop new strains of plant that are more disease resistant, can grow in different climates, and can give greater yields.

This can't happen fast enough. According to data in a report issued by the Club of Rome in 1972 (D. H. Meadows, D. L. Meadows, J. Randers, and W. Behrens III, *The Limits of Growth*), with the then existing crop strains and farming methods, the total area of agriculturally-viable land on the earth's surface would cease to be sufficient to feed the world's population in the year 2,000. Though distribution problems mean that we still see famine in several parts of the world, there is currently no overall food shortage. The development of new crop strains and improved farming methods has prevented that 1972 prediction from being realized. But with a world population still undergoing exponential growth with a ratio close to 2, it will take further advances to ensure that the world has sufficient food to survive the coming half century, after which the world population is—according to the latest population models—predicted to enter a period of decline.

Thus, it is not at all unreasonable to claim that recreational mathematics could lead to—and might in fact have already played a part in—feeding a growing world population.

Breeding Rabbits in a Chancy World

Of course, growing systems don't always exactly follow mathematically precise patterns such as the Fibonacci sequence. The real world is full of random events and unpredictable changes. Genuine rabbit populations tend to grow somewhat erratically, and if you go around counting petals and leaves of flowers and plants you won't always find Fibonacci numbers. (In fact, many flowers have six petals, but that's because of the biological efficiency of six-fold symmetry, so recreational mathematics still plays a role.)

Can mathematicians say anything about Fibonacci-like growth sequences that are subjected to a random effect? Surprisingly perhaps, the answer is yes, under certain circumstances.

As I mentioned earlier, as you proceed along the Fibonacci sequence, the ratios of the successive terms get closer and closer to the golden ratio, $1.61803\ldots$ Another way to express the same result is that the Nth Fibonacci number is approximately equal to a constant times the Nth power of the golden ratio. This gives a way to calculate the Nth Fibonacci number without generating the entire sequence of preceding Fibonacci numbers: Take the golden ratio, raise it to the power N, divide by the square root of 5, and round off the result to the nearest whole number. The answer you get will be the Nth Fibonacci number.

Faced with such a result, most numerically minded citizens will nod appreciatively and move on to something else. But mathematicians—both professional and recreational—ask "What if?" questions. For example, suppose that, when you generate the Fibonacci sequence, you flip a coin at each stage. If it comes up heads, you add the last number to the one before it to give the next number, just as Fibonacci did. But if it comes up tails, you subtract. Now you have a Fibonacci-like process in which chance plays a role.

For example, one possible sequence you could get in this way is:

1, 1, 2 (H), 3 (H), −1 (T), 4(T), −5 (T), −1 (H), …

Another is:

1, 1, 0 (T), 1 (H), −1 (T), 2 (T), 1 (H), 1 (T), …

The random sign changes can lead to sequences that suddenly switch from large positive to large negative, such as:

1, 1, 2, 3, 5, 8, 13, 21, −8, 13, −21, …

as well as to sequences that cycle endlessly through a particular pattern, such as:

1, 1, 0, 1, 1, 0, 1, 1, 0, 1, 1, 0, …

or

1, 1, 0, 1, −1, 0, −1, 1, 0, 1, −1, …

If you are like the character Rosenkrantz in Tom Stoppard's play *Rosenkrantz and Guildenstern Are Dead* and your coin keeps coming up heads every time, you can even get the original Fibonacci sequence

1, 1, 2, 3, 5, 8, 13, 21, 34, 55, 89, …

With such a variety of behavior, it's not obvious that such sequences follow the nice kind of growth pattern of the Fibonacci sequence.

But they do. In 1998, a young mathematician called Divakar Viswanath showed that the absolute value of the Nth number in any random Fibonacci sequence generated as described is approximately equal to the Nth power of the number 1.13198824…

Actually, that's not quite accurate. Because the sequences are generated randomly, there are infinitely many possibilities. Some of them will not have the 1.13198824 property. For example, the sequence that cycles endlessly through 1, 1, 0 does not have the property, nor does the original Fibonacci sequence. But those are special cases. What Viswanath showed is that if you actually start to generate such a sequence, then with probability 1 the sequence you get will have the 1.13198824 property. In other words, you can safely bet your life on the fact that for your sequence, the bigger N is, the closer the absolute value of the Nth number gets to the Nth power of 1.13198824…

To give some idea of what this result says, the way the randomized Fibonacci sequence is generated is a bit like the daily weather at a particular location. Today's weather can be assumed to depend on the weather the previous two days, but there is a large element of chance.

The analog of the number 1.13198824... for the weather would give a quantitative measure of the unpredictability of weather. It measures the rate at which small disturbances explode exponentially in time. It would tell you for exactly how many days high-speed computers can forecast weather reliably. Unfortunately, nobody knows this number for global weather, and probably never will.

Viswanath's result brought to an end a puzzle that had its origins in 1960. In that year, Hillel Furstenberg (now at the Hebrew University) and Harry Kesten (at Cornell University) showed that for a general class of random-sequence generating processes that includes the random Fibonacci sequence, the absolute value of the Nth member of the sequence will, with probability 1, get closer to the Nth power of some fixed number. (The exact formulation of their result is in terms of random matrix products, and is definitely not recreational mathematics. See Viswanath's paper (published in the journal *Mathematics of Computation*) for an exact statement, or read the whole story in the book *Random Products of Matrices With Applications to Infinite-Dimensional Schrödinger Operators*, by P. Bougerol and J. Lacroix, published by Birkhäuser, Basel, in 1984.)

Since Furstenberg and Kesten's deep result applied to the randomized Fibonacci process, it followed that, with probability 1, the absolute value of the Nth number in any random Fibonacci sequence will get closer and closer to the Nth power of some fixed number K. But no one knew the value of the number K, or even how to calculate it. What Viswanath did was find a way to compute K. At least, he computed the first eight decimal places. Almost certainly, K is irrational, so cannot be computed exactly.

Since there is no known algorithm to compute K, Viswanath had to adopt a circuitous route, showing that K equals e^P, where P lies somewhere between 0.1239755980 and 0.1239755995 (and, as usual, e is the base for natural logarithms). Since those two numbers are equal in their first eight decimal places, that meant he could calculate K to eight decimal places.

The process involved large doses of mathematics and some heavy duty computing. Since his computation made use of floating point arithmetic—which is not exact—Viswanath had to carry out a detailed mathematical analysis to obtain an upper bound on any possible errors in the computation.

He described the key to his new result this way: "The problem was that fractals were coming in the way of an exact analysis. What I did was to guess the fractal and use it to find K. To do this, I made use of some devilishly clever work carried out by Furstenberg in the early 1960s."

And with that computation, mathematics had a new constant, a direct descendent of a pair of rabbits in thirteenth-century Italy.

From Rabbits to How We Can See through Glass

Does Viswanath's new result have any applications? Probably not—unless you count the fact that an easily understood, cute, counter-intuitive result about elementary integer arithmetic can motivate a great many individuals (your present reporter included) to take a look at an area of advanced mathematics full of deep and fascinating results that has perhaps not hitherto had the attention it deserves. Indeed, in that respect, the randomized Fibonacci sequence problem resembles Fermat's Last Theorem, finally solved by Andrew Wiles in 1994. It too was easy to state and to understand, and yet it was only a hairsbreadth away from some of the deepest and most profound mathematics of all time. Over the years, Fermat's Last Theorem attracted many people—both professional and amateur—to learn about analytic number theory (including Wiles himself). When we talk about "applications," we often overlook the very important application of attracting people to mathematics in the first place.

If it's "real" applications you want, however—perhaps with a chance of saving the world—then you don't have to go any further than the work of Furstenberg and Kesten that lays behind Viswanath's result. Applications of that work have led to advances in lasers, new industrial uses of glass, and to development of the copper spirals used in birth control devices. The research which led to those advances earned the 1977 Nobel Prize in physics for the three individuals involved: Philip Anderson of Bell Laboratories, Sir Neville Mott of Cambridge University in England, and John van Vleck of Harvard.

The citation that accompanied the Nobel Prize to the three researchers declared it to be "for their fundamental theoretical investigations of the electronic structure of magnetic and disordered systems."

"Disordered systems" exists within noncrystallic materials that have irregular atomic structures, making it difficult to theorize about them. The key starting point for their work was to realize the importance of electron correlation—the interaction between the motions of the electrons.

Anderson's main contribution was the discovery a phenomenon known as Anderson localization, and this is where the random matrix multiplication came in. Imagine you have a material, say a semiconductor, with some impurities. If you pass a current through it, you might expect it to get dispersed and diffracted in a random fashion by the impurities. But in fact, at certain energies, it stays localized. The first rigorous explanation of this used Furstenberg and Kesten's work.

A similar explanation shows why you can see through glass. The irregular molecular structure of glass—technically it's a liquid—should surely cause some of the incident light rays to bounce around in a seemingly random fashion, resulting in a blurred emergent image. But as we all know, that's not what happens. Somehow, the repeated random movements lead to orderly behavior. Furstenberg and Kesten's work on random matrix multiplication provides the mathematical machinery required to explain how this happens.

Of course, with the Furstenberg and Kesten work, we are no longer talking about recreational mathematics. But it does provide another excellent illustration of how a piece of "curiosity driven, pure mathematical research" can, years later, lead to extremely useful applications. And is so doing it reminds us that the pursuit of mathematics purely for fun—something Martin encouraged millions of people to do for many years—can sometimes have important consequences that affect us all.

Bibliography

[BL84] P. Bougerol and J. Lacroix. *Random Products of Matrices With Applications to Infinite-Dimensional Schrödinger Operators*. Birkhauser, Basel, 1984.

[FK60] Hillel Furstenberg and Harry Kesten. Products of random matrices. *Annals of Mathematical Statistics*, 31:457–469, 1960.

[MMRI72] D. H. Meadows, D. L. Meadows, J. Randers, and W. Behrens III. *The Limits of Growth*. Universe Books, New York, 1972. Reported issued by *Club of Rome* in 1972.

[Vis00] Divakar Viswanath. Random fibonacci sequences and the number 1.13198824.... *Mathematics of Computation*, 69(231):1131–1155, 2000.

Variations on a Transcendental Theme

Roger Penrose

The simple ideas that I am describing here represent the minutest token of the appreciation and respect that I have for Martin Gardner. His influence has been immeasurable, certainly on me personally when I was an aspiring young mathematician, but similarly on a great many others in developing their interest and excitement for mathematical ideas.

From some time in the 1970s, I have used the following unending sequence of numbers

$$\ldots, 7, 9, 12, ?, 24, 36, 56, 90, \ldots$$

as an example of a deceptively simple-looking puzzle, which hides a certain subtlety. The puzzle is to supply the missing number, denoted by ?. Of course, one could simply use a sixth degree polynomial to yield the numbers that are explicitly given, and this provides the answer

$$? = \frac{582}{35} \approx 16.629$$

(easily obtained by taking differences repeatedly). But this is not the intended answer, which is numerically only slightly larger

$$? = 24 \log 2 \approx 16.636$$

(all logarithms, in this article, being natural logarithms).

Why is this second suggestion actually the "correct answer"? The idea is that there should be a *simple* formula to represent all the terms

---

Roger Penrose was knighted in 1994 for his numerous contributions to mathematics and science.

of the infinite sequence, where the *zeroth* term is that given by "?". In this case the formula for the nth term is

$$\frac{24(2^n - 1)}{n},$$

and we easily check that the values for $n = -3, -2, -1, +1, +2, +3, +4$ come out correctly. The value for $n = 0$ appears to be the indeterminate "0/0", but we obtain the required answer by taking the limit $n \to 0$, (using l'Hôpital's rule and $2^n = e^{n \log 2}$).

The thing that is striking about this is that, *except for $n = 0$*, the formula always yields a *rational number* for each integer value of n. Yet for $n = 0$ the formula actually provides a *transcendental* value. (I came to this example whilst worrying about a problem of relevance to physics having this kind of character.) In this example, I have disguised the fact that the general value is merely rational, rather than being integral, by choosing a multiplying factor (here 24) that eliminates all the visible denominators. Including, a few further terms perhaps slightly spoils the effect:

$$\ldots, 5\tfrac{5}{8}, 7, 9, 12, ?, 24, 36, 56, 90, 148\tfrac{4}{5}, 252, 435\tfrac{3}{7}, 765, \ldots!$$

It is not hard to construct other examples with the same property, but the numbers often get unreasonably large. We can take our $(2^n - 1)/n$ the "other way up", the general term now being $105n/(1 - 2^{-n})$, where we find

$$\ldots, 28, 45, 70, 105, ?, 210, 280, 360, 448, \ldots$$

for which the answer is $? = 105/\log 2 \approx 151.48$. For another example, we find the symmetrical

$$\ldots, 85, 56, 40, 32, ?, 32, 40, 56, 85, \ldots$$

where the general term is $64(2^n - 2^{-n})/3n$ and

$$? \approx 29.57.$$

A little more exotic are examples for which a fair case can be made that "?" should most naturally taken to be a complex number. For example,

$$\ldots, 16, -12, 9, -8, 6, -12, ?, 12, 6, 8, 9, 12, 16, \ldots$$

is given by the expression $12F_n/n$, where F_n is the nth Fibonacci number:

n	$\ldots$	-4	-3	-2	-1	0	1	2	3	4	5	$\ldots$
F_n	$\ldots$	-3	2	-1	1	0	1	1	2	3	5	$\ldots$

satisfying

$$F_n = F_{n-1} + F_{n-2}.$$

Fibonacci numbers can be defined analytically by use of the formula

$$F_n = \frac{\phi^n - (-\phi)^{-n}}{\sqrt{5}}$$

where ϕ is the golden mean

$$\phi = \frac{1 + \sqrt{5}}{2}.$$

Using the same procedure as above for obtaining the value of our expression $12F_n/n$ at $n = 0$, and making use of

$$\phi^n - (-\phi)^{-n} = e^{n \log \phi} - e^{-n \log(-\phi)}$$

and

$$\log(-\phi) = i\pi + \log \phi$$

so that l'Hôpital's rule gives us

$$\lim_{n \to 0} \frac{\phi^n - (-\phi)^{-n}}{n} = \log \phi + i\pi + \log \phi$$

we find

$$\lim_{n \to 0} \frac{F_n}{n} = \frac{2 \log \phi + i\pi}{\sqrt{5}}$$

so we obtain the complex transcendental answer

$$? = \frac{24}{\sqrt{5}} \log \phi + \frac{12\pi i}{\sqrt{5}} \approx 5.165 + i16.86.$$

Of course, there are conventional matters involved in the choice of the logarithm of a complex number, and we could alternatively have come up with $? = (24 \log \phi - 12\pi i)/\sqrt{5}$. (There is no algebraic way of distinguishing $+i$ from $-i$, for example.) Some people might prefer the average of these two answers, which drops the imaginary part altogether, giving simply $? = 24 \log \phi/\sqrt{5}$. My own opinion is that it is more natural to choose the complex-value $? = (24 \log \phi + 12\pi i)/\sqrt{5}$, as above—or even any of the other values $? = (24 \log \phi + 12\pi i + 24N\pi i)/\sqrt{5}$, where N is some integer—than it is to adopt $? = 24 \log \phi/\sqrt{5}$. (My reasons perhaps come from an interest in quantum field theory, where things of this nature are indeed often appropriate.)

In view of such ambiguities, it is perhaps more appropriate to phrase such puzzles in the form "find a mathematical justification for the choice $? = (24 \log \phi + 12\pi i)/\sqrt{5}$ in the sequence $\ldots, 16, -12, 9, -8, 6, -12, ?, 12, 6, 8, 9, 12, 16, \ldots$" In this spirit, we can address other examples of this nature. One of the simplest is just the "upside-down" version of the example just considered, for which we need to justify

$$? = \frac{24\sqrt{5}(2\log \phi - i\pi)}{4(\log \phi)^2 + \pi^2} \approx 11.549 + i5.228$$

in the sequence

$$\ldots, 9, -12, 16, -18, 24, -12, ?, 12, 24, 18, 16, 12, 9, \ldots.$$

For three more somewhat similar sequences, consider

$$\ldots, -246, 153, -104, 72, -96, \quad ?, \quad 96, 72, 104, 153, 246, \ldots$$
$$\ldots, 5, -8, 8, -32, \quad ?, \quad 64, 32, 64, 80, \ldots$$
$$\ldots -2, \quad ?, \quad 12, 6, 28, 69, 132, 266, \ldots$$

where one has to justify, respectively,

$$? = \frac{192}{5}(2\log 2 + i\pi)$$

$$? = \frac{64}{3}(\log 2 - i\pi)$$

$$? = \frac{12}{5}\left(\log \frac{3}{2} - i\pi\right).$$

Perhaps some reader will come up with an original further class of examples.

Bibliography

[Pen] R. Penrose. Articles in *Twistor Newsletter*, an informal publication of the Mathematical Institute, 24–29 St Giles, Oxford OX1 3LB, UK. Issues **10**, p. 22, July 1980; **41**, p. 37, August 1996; **42**, p. 25, March 1997; and [erratum **43**, p. 34, July 1980].

Magic "Squares" Indeed!

Arthur T. Benjamin and Kan Yasuda

Introduction

Behold the remarkable property of the magic square:

$$\begin{bmatrix} 6 & 1 & 8 \\ 7 & 5 & 3 \\ 2 & 9 & 4 \end{bmatrix}$$

$$
\begin{aligned}
618^2 + 753^2 + 294^2 &= 816^2 + 357^2 + 492^2 \text{ (rows)} \\
672^2 + 159^2 + 834^2 &= 276^2 + 951^2 + 438^2 \text{ (columns)} \\
654^2 + 132^2 + 879^2 &= 456^2 + 231^2 + 978^2 \text{ (diagonals)} \\
639^2 + 174^2 + 852^2 &= 936^2 + 471^2 + 258^2 \text{ (counter-diagonals)} \\
654^2 + 798^2 + 213^2 &= 456^2 + 897^2 + 312^2 \text{ (diagonals)} \\
693^2 + 714^2 + 258^2 &= 396^2 + 417^2 + 852^2 \text{ (counter-diagonals)}.
\end{aligned}
$$

This property was discovered by Dr. Irving Joshua Matrix [Gar89], first published in [Hol70] and more recently in [Bar97]. We prove that this property holds for *every* 3-by-3 magic square, where the rows, columns, diagonals, and counter-diagonals can be read as 3-digit numbers in *any* base. We also describe n-by-n matrices that satisfy this

Mathemagician **Arthur T. Benjamin** teaches mathematics at Harvey Mudd College in Claremont, California and performs lightning calculations at the Magic Castle in Hollywood. **Kan Yasuda** is a graduate student studying mathematics at the University of Tokyo. This article was reprinted from the *American Mathematical Monthly* with permission from the MAA and the authors.

365

condition, among them all circulant matrices and all symmetrical magic squares. For example, the 5-by-5 magic square in (1) also satisfies the square-palindromic property for every base.

$$\begin{bmatrix} 17 & 24 & 1 & 8 & 15 \\ 23 & 5 & 7 & 14 & 16 \\ 4 & 6 & 13 & 20 & 22 \\ 10 & 12 & 19 & 21 & 3 \\ 11 & 18 & 25 & 2 & 9 \end{bmatrix} \tag{1}$$

We must be careful when we read these numbers. The base 10 number represented by the first row of (1) is $17 \cdot 10^4 + 24 \cdot 10^3 + 1 \cdot 10^2 + 8 \cdot 10 + 15 = 194195$. The base 10 number based on the first row's reversal is 158357.

Sufficient Conditions

We say that a real matrix is *square-palindromic* if, for every base b, the sum of the squares of its rows, columns, and four sets of diagonals (as in the previous examples) are unchanged when the numbers are read "backwards" in base b. We can express this condition using matrix notation. Let M be an n-by-n matrix. Then the n numbers (in base b) represented by the rows of M are the entries of the vector $M\mathbf{b}$, where $\mathbf{b} = (b^{n-1}, b^{n-2}, \ldots, b, 1)^T$, and T denotes the transpose operation. The sum of the squares of these numbers is

$$(M\mathbf{b})^T (M\mathbf{b}) = \mathbf{b}^T (M^T M)\mathbf{b}.$$

Next, the n numbers represented by the rows when read "backwards" are the entries of $MR\mathbf{b}$ where the n-by-n *reversal matrix* $R = [r_{ij}]$ has $r_{ij} = 1$ if $i + j = n + 1$, and $r_{ij} = 0$ otherwise. Note that $R^T = R^{-1} = R$. The sum of the squares of these numbers is

$$(MR\mathbf{b})^T (MR\mathbf{b}) = \mathbf{b}^T (R(M^T M)R)\mathbf{b}.$$

Hence a sufficient condition for the *rows* of M to satisfy the square-palindromic property is simply $R(M^T M)R = M^T M$. Matrices A that satisfy $RAR = A$ are called *centro-symmetric*, [Wea85]: $a_{ij} = a_{n+1-i,n+1-j}$. Matrices A that satisfy $RAR = A^T$ are called *persymmetric*, [GL83]: $a_{ij} = a_{n+1-j,n+1-i}$. It is easy to see that symmetric matrices that are centro-symmetric must also be persymmetric. Since $M^T M$ is necessarily symmetric, our sufficient condition says that $M^T M$ is centro-symmetric, or equivalently, that

$$M^T M \text{ is persymmetric.}$$

The square-palindromic condition for the *columns* of M is the square-palindromic condition for the rows of M^T. Hence it suffices to require that

$$MM^T \text{ is persymmetric.}$$

For the first set of *diagonals*, we create a matrix $\tilde{M}$ with the property that each column of $\tilde{M}$ represents a diagonal starting from the first row of M. To do this, we introduce two other special square matrices. Let $P_k = [p_{ij}]$ denote the n-by-n *projection matrix* whose only non-zero entry is $p_{kk} = 1$. Notice that $P^T = P$, and $P_k M$ preserves the k-th row of M but turns all other rows to zeros. Let $S = [s_{ij}]$ denote the n-by-n *shift operator* where $s_{ij} = 1$ if $i - j \equiv 1 \pmod{n}$, $s_{ij} = 0$ otherwise.

The following properties of S are easily verified: $S^n = I_n$, $S^{-1} = S^T = RSR$, and MS^k shifts the columns of M over "k steps to the left". Now define

$$\tilde{M} = \sum_{i=1}^{n} P_i M S^{i-1}.$$

Hence the i-th diagonal of M, starting from the first row becomes the i-th column of $\tilde{M}$. By the column condition, these diagonals satisfy the square-palindromic property if the (i, j) entry of $\tilde{M}\tilde{M}^T$ equals its $(n + 1 - j, n + 1 - i)$ entry.

We have

$$\tilde{M}\tilde{M}^T = \sum_{i=1}^{n} P_i M S^{i-1} \left(\sum_{j=1}^{n} P_j M S^{j-1} \right)^T$$

$$= \sum_{i=1}^{n} \sum_{j=1}^{n} P_i M S^{i-j} M^T P_j.$$

It follows that $\tilde{M}\tilde{M}^T$ has the same (i, j) entry as $MS^{i-j}M^T$, and the same $(n + 1 - j, n + 1 - i)$ entry as well; if $MS^{i-j}M^T$ is persymmetric, then these entries are equal. Consequently, these diagonals obey the square-palindromic property if

$$MS^k M^T \text{ is persymmetric for } k = 1, \ldots, n. \tag{2}$$

Conveniently, (2) also ensures that the counter-diagonals starting from the first row satisfy the square-palindromic property. This can be seen by mimicking the preceding explanation with $\tilde{M} = \sum_{i=1}^{n} P_i M S^{-(i-1)}$, whereby $\tilde{M}\tilde{M}^T$ has the same (i, j) and $(n + 1 - j, n + 1 - i)$ entry as $MS^{j-i}M^T$. For the other diagonal and counterdiagonal, we obtain similar results [Yas97], which we summarize in the following theorem:

Theorem 1 *A square matrix M has the square-palindromic property if the following matrices are all persymmetric:*

1. $M^T M$,

2. MM^T,

3. $MS^k M^T$, *for $k = 1, \ldots, n$, and*

4. $M^T S^k M$, *for $k = 1, \ldots, n$.*

Square-Palindromic Matrices

Next we explore classes of matrices that are square-palindromic. We say that a square matrix A is *centro-skew-symmetric* if $RAR = -A$, that is, $a_{ij} + a_{n+1-i,n+1-j} = 0$.

$$\begin{bmatrix} 1 & 2 & 3 & 4 \\ 5 & 6 & 7 & 8 \\ 8 & 7 & 6 & 5 \\ 4 & 3 & 2 & 1 \end{bmatrix} \qquad\qquad \begin{bmatrix} a & b & c \\ d & 0 & -d \\ -c & -b & -a \end{bmatrix}$$

Centro-Symmetric   Centro-Skew-Symmetric

Theorem 2 *Every centro-symmetric or centro-skew-symmetric matrix is square-palindromic.*

Proof: If M is centro-symmetric or centro-skew-symmetric, then the relations $RM = \pm MR$ and $R(S^k)R = S^{-k}$ ensure that M satisfies the conditions of Theorem 1.   □

The theorem is not at all surprising since the collection of rows, columns and diagonals of M read the same backwards and forwards. The next class of matrices, however, satisfies the conditions in a non-obvious way.

We say that A is *circulant* if every entry of each "diagonal" is the same, i.e., $a_{ij} = a_{k\ell}$ if $i - j \equiv k - \ell \bmod n$, or simply $SAS^{-1} = A$. We say that A is *(-1)-circulant* if $SAS = A$.

$$\begin{bmatrix} 1 & 2 & 3 & 4 \\ 4 & 1 & 2 & 3 \\ 3 & 4 & 1 & 2 \\ 2 & 3 & 4 & 1 \end{bmatrix} \qquad \begin{bmatrix} 1 & 2 & 3 & 4 & 5 \\ 2 & 3 & 4 & 5 & 1 \\ 3 & 4 & 5 & 1 & 2 \\ 4 & 5 & 1 & 2 & 3 \\ 5 & 1 & 2 & 3 & 4 \end{bmatrix}$$

Circulant   (−1)-Circulant

Notice that the circulant and (−1)-circulant property is preserved under transposing. It is easy to show that the product of two circulant matrices or two (−1)-circulant matrices is circulant, while the product of a circulant and (−1)-circulant matrix is (−1)-circulant. Note that S is circulant, R is (−1)-circulant, and that all circulant matrices are persymmetric since a_{ij} and $a_{n+1-j,n+1-i}$ lie on the same diagonal. Consequently, if M is circulant or (−1)-circulant, the matrices $M^T M$, MM^T, $MS^k M^T$, and $M^T S^k M$ are all circulant, and thus persymmetric. From Theorem 1, it follows that

Theorem 3 *Every circulant or (−1)-circulant matrix is square-palindromic.*

Notice that four of the six square-palindromic identities are not obvious, but two of the diagonal sums are completely trivial!

Magic and Semi-Magic Squares

A *semi-magic square* with magic constant c is a square matrix A in which every row and column adds to c. Using matrix notation, this says that $AJ = cJ = JA$, where J is the matrix of all ones. If the main diagonal and main counter-diagonal also add to c, then the matrix is called a *magic square*. Circulant and (-1)-circulant matrices are always semi-magic, but are not necessarily magic.

A magic square A is *symmetrical* [BJ76] if the sum of each pair of two entries that are opposite with respect to the center is $2c/n$, that is $a_{ij} + a_{n+1-i,n+1-j} = 2c/n$. Notice that a semi-magic square with this property is magic.

Like the example below, magic and semi-magic squares do not necessarily satisfy the square-palindromic property.

$$\begin{bmatrix} 2 & 0 & 1 \\ 0 & 2 & 1 \\ 1 & 1 & 1 \end{bmatrix}$$

Semi-Magic but not square-palindromic

However,

Theorem 4 *Every symmetrical magic square is square-palindromic.*

Proof: The trick is to notice that if M is a symmetrical magic square with magic constant c, then $M = M_0 + cJ/n$, where M_0 is a symmetrical magic square with magic constant 0. But this implies that M_0 is centro-skew-symmetric. Therefore M_0 is square-palindromic and satisfies the conditions of Theorem 1. Thus, since $M_0^T M_0$ and J are persymmetric, it follows that $M^T M = (M_0 + cJ/n)^T(M_0 + cJ/n) = M_0^T M_0 + c^2 J/n$ is also persymmetric. Hence M satisfies condition 1 of Theorem 1. To verify condition 3 (the other cases are similar), notice that

$$
\begin{aligned}
MS^k M^T &= (M_0 + \frac{c}{n}J)S^k(M_0 + \frac{c}{n}J)^T \\
&= M_0 S^k M_0^T + \frac{c^2}{n}J
\end{aligned}
$$

is persymmetric for $k = 1, \ldots, n$, since M_0 satisfies condition 3 of Theorem 1.

Although not all magic squares are square-palindromic, it is easy to see that all 3-by-3 magic squares are symmetrical. Consequently, we have

Theorem 5 *All 3-by-3 magic squares are square-palindromic.*

Bibliography

[Bar97] E. J. Barbeau. *Power Play*. Mathematical Association of America, Spectrum, Washington, D.C., 1997.

[BJ76] W. H. Benson and O. Jacoby. *New Recreations with Magic Squares*. Dover Publications, New York, 1976.

[Gar89] M. Gardner. *Penrose Tiles to Trapdoor Ciphers... And the Return of Dr. Matrix*. W. H. Freeman and Company, 1989.

[GL83] G. H. Golub and C. F. Van Loan. *Matrix Computations*. Johns Hopkins University Press, 1983.

[Hol70] R. Holmes. The magic magic square. *The Mathematical Gazette*, 54(390):376, 1970.

[Wea85] J. R. Weaver. Centro-symmetric (cross-symmetric) matrices, their basic properties, eigenvalues, and eigenvectors. *The American Mathematical Monthly*, 92(10):711–717, 1985.

[Yas97] K. Yasuda. A square sum property of magic squares. Senior Thesis, Mathematics Department, Harvey Mudd College, 1997.

The Beer Bottles Problem

N. G. de Bruijn

In order to have a party on the top of a cliff, two people, P_1 and P_2, want to transport a large number of beer bottles from the ground level to the top. There are two paths, both with several ups and downs. The bottles are all placed on a rigid board of sufficient length, and the idea is that P_1 takes the first path, P_2 the second one. But since the board is slippery, it should be kept absolutely horizontal all the time, for otherwise the bottles will slide down. Can this always be prevented?

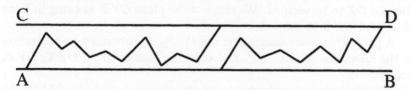

Figure 1. The two paths.

The situation is shown in Figure 1. A point P_1 has to move on the first path, P_2 on the second one. P_1 and P_2 start on the ground level AB and have to finish at the top level CD in such a way that they had the same altitude at each moment in-between. This is indeed possible, at least if we assume that the paths lie entirely between AB and CD, and that they do not have infinitely many maxima and minima. Let us just suppose that they are composed of a finite number of straight line segments, as in Figure 1.

This note will show that the problem is solvable indeed, and it will use old-fashioned descriptive geometry. The notation is shown in Figure 2.

N. G. de Bruijn has published papers since 1937 on a wide variety of subjects in mathematics and related areas, and maintains a keen interest in mathematical entertainment.

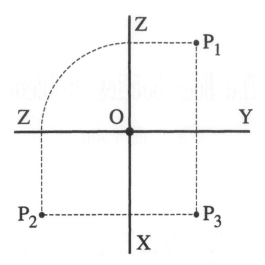

Figure 2. Three projections folded into a single plane.

We have three mutually orthogonal lines OX, OY, and OZ in three-dimensional space. Let us imagine the plane OXY to be horizontal, and the axis OZ to be vertical. We think of the plane OYZ as being in front of us, and OXZ to our left.

A point P in space leads to points P_1, P_2, P_3, obtained by projection in the directions of OX, OY, OZ onto the planes OYZ, OZX, OYZ, respectively.

The projections are shown in one and the same plane by folding the plane OXZ down, folding along the line OX, and, similarly, folding the plane OYZ down along OY. As long as we have points in the first octant (i.e., in front of OYZ, to the right of OZX, and above OXY), the projections do not overlap.

Now we put the first path of Figure 1 on the plane OYZ. It will act as the orthogonal projection of a factory roof. That roof is formed by taking, through each point of the path, a straight line parallel to OX. Similarly we put the second path on the plane OXZ, and taking lines parallel to OY, we get a second roof.

The *intersection* of the two roofs is significant. If a point P lies on the intersection, then its projection P_1 lies on the first path, and P_2 on the second one. Moreover, these projections P_1 and P_2 have the same altitude, simply the altitude of P above the horizontal plane. In other words, P represents an acceptable situation of the two people P_1 and P_2, acceptable in the sense that the bottles do not begin to slide.

The solution of the problem of the beer bottles can now be obtained by the construction of the intersection of the roofs by means of the methods of descriptive geometry. What we have to construct is the projection of the intersection on the horizontal plane, and what we have to show is that this intersection contains a continuous connection from the upper left corner O (representing the beginning of the paths) to the lower right corner E (corresponding to the upper end of the paths). And this connectivity can be studied just by looking at the projection on the horizontal plane OXY.

We shall not go into the details of the construction, but just display the result in the example of Figure 3. One arbitrary position of the point P is indicated by its projections P_1, P_2, P_3. In the lower right corner, between the lines OX and OY we have a graph drawn in heavy lines. Indeed it is possible to walk through this graph from point O in the upper left corner to point E in the lower right. This means that the beer gets to the party all right.

Let us now argue why there is always such a connection. It follows from a simple theorem by Euler. It says that if in a finite (undirected) graph where each vertex has an even degree (the degree is the number

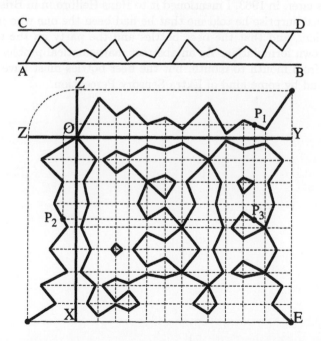

Figure 3.   (Bottom) The intersection of the roofs. (Top) The two paths.

of connections to other vertices), then the set of edges can be split into a number of disjoint cycles. As a corollary we mention that if all vertices have even degree apart from the vertices O and E whose degree is odd, then there is a connection from O to E through the graph. In order to reduce it to Euler's theorem, we just add a single edge from O to E, getting a graph where all degrees are even, next split the graph into cycles, then take the cycle containing OE. Removing OE from that cycle, we have a path from O to E.

In our case of the graph obtained as horizontal projection of the intersection of two roofs, each vertex has degree 0, 2, or 4, except for the points O and E, which have just 1. This proves that O and E are connected through the graph.

This descriptive geometry method is not just a proof of the existence of a solution: it also provides a quick survey of all possibilities to walk from O to E.

Figure 3 was produced by a computer program (in PostScript) that gets the data of the two paths as input.

A personal note: In a mathematical formulation the problem went around in the fifties. At that time I gave a talk about it in a students club, and there I invented the story of the party with the beer bottles. A few years later, in 1963, I mentioned it to Hans Heilbronn in Bristol, but to my great surprise he told me that he had been the one who invented the problem, and that the beer bottles and the party on the cliff had been his own formulation. As an abstract mathematical problem it had traveled from mouth to mouth, but the beer bottles must have arrived in my mind by some kind of Extra Sensory Perception.

The Fractal Society

Clifford Pickover

> "If the cosmos were suddenly frozen, and all movement ceased. a
> survey of its structure would not reveal a random distribution of
> parts. Simple geometrical patterns, for example, would be found
> in profusion—from the spirals of galaxies to the hexagonal shapes
> of snow crystals. Set the clockwork going, and its parts move
> rhythmically to laws that often can be expressed by equation of
> surprising simplicity. And there is no logical or *a priori* reason
> why these things should be so."
>
> —Martin Gardner, *Order and Surprise*

Imagine a group of mathematicians who meet each month in a secret club. Status in their "Fractal Society" is based on the prowess
with which an individual plays mathematical games and proves mathematical theorems. The center of such activity is a building called the
Imaginarium, which is shaped like a Mandelbrot set. There are various
pleasurable rewards bestowed upon club members in proportion to the
novelty of theorems they solve. A favorite society game is called Fractal
Fantasies.

The playing board for the Fractal Fantasies game is a fractal nesting of interconnected rectangles (Figure 1). The Fractal Society is so
enthralled with this game that they have cut the design into the roofing
slabs of their homes and in the surfaces of their kitchen tables. The
board for Fractal Fantasies contains rectangles within rectangles interconnected with gray lines as shown in Figure 1. There are always two
rectangles within the rectangles that encompass them. The degree of

Clifford Pickover is a leading popular science writer whose latest book is
Wonders of Numbers: Adventures in Mathematics, Mind, and Meaning. See
http://www.pickover.com.

nesting can be varied. Beginners play with only a few nested rectangles, while grand masters play with many recursively positioned rectangles. Tournaments last for days, with breaks only for eating and sleeping.

The playing board illustrated in Figure 1 is called a "degree 2" board, because it has two different sizes of rectangles within the large bounding rectangle. The "degree 3" board is shown in Figure 2. Beginners usually start with a degree 1 board, and grand masters have been known to use a degree 20 board. One player uses white playing pieces (like stones), the other uses black. Each player starts with a number of pieces equal to half the number of vertices (small circles) on the board minus two. For the board here, each player gets 19 stones. With alternate moves, the players begin by placing a stone at points on the circles which are empty. As they place stones, each player attempts to form a row of three stones along any one of the horizontal sides of any rectangle. This three-in-a-row assembly of stones is called a Googol. When all the stones have been placed, players take turns moving a piece to a neighboring vacant space along one of the dashed or straight connecting lines. When a player succeeds in forming a Googol (either during the alternate placement of pieces at the beginning of the game, or during alternate moves along lines to adjacent empty points) then the player captures any one of the opponent's pieces on the board and removes it from the board. These removed stones may be kept in ✿-shaped receptacles on opposing sides of the board in Figure 1. (In some versions of the game, an opposing stone cannot be taken from an opposing Googol.) A player loses when he or she no longer has any pieces or cannot make a move.

Mathematicians and philosophers will no doubt spend many years pondering a range of questions, particularly for boards with higher nesting. Computer programmers will design programs allowing the board to be magnified in different areas permitting the convenient playing at different size scales. They'll all wish they had fractal consciousnesses allowing the contemplation of all levels of the game simultaneously.

Many of obsessed friends have spent years of their lives pondering the following questions relating to Fractal Fantasies. No one has succeeded in answering these questions for games with degree higher than 2. Various centers have been established and funded in order to answer the following research questions:

1. What is the maximum number of pieces that can be on the board without any forming a row?

2. Is there a best opening move?

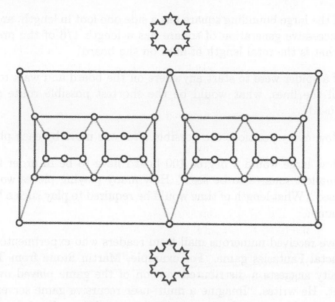

Figure 1.   The degree 2 game.

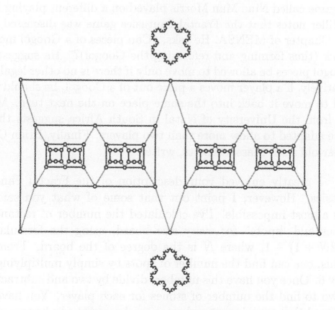

Figure 2.   The degree 3 game.

3. If the large bounding square has a side one foot in length, and each successive generation of square has a length 1/6 of the previous, what is the total length of lines on the board?

4. If a spider were to start anywhere on the board and walk to cover all the lines, what would be the shortest possible route on the board?

5. How many positions are possible after one move by each player?

6. How large would a degree 100 board have to be in order for the smallest squares to be seen? How many playing pieces would be used? What length of time would be required to play such a bizarre game?

I have received numerous mail from readers who experimented with the Fractal Fantasies game. For example, Martin Stone from Temple University suggests a distributed version of the game played over the Internet. He writes, "Imagine a multi-user recursive game server dedicated to the fostering of a greater intuitive understanding of recursive structures and permutations." David Kaplan from New York University points out that the game rules for Fractal Fantasies are similar to a medieval game called Nine Man Morris played on a different playing board. Paul Miller notes that the Fractal Fantasies game was discussed at the Boston Chapter of MENSA. He asks, "Can pieces of a Googol move out and back (thus forming and reforming the Googol)?" He suggests that the Googol pieces be allowed to move only if there is no other legal move. Alternatively, if a player moves a piece out of a Googol, he should not be allowed to move it back into the same place on the next turn. Michael Currin from the University of Natal in South Africa suggests that the game be adapted to allow more than two players. Finally, Brian Osman, a 15-year-old from Massachusetts, writes:

> I greatly enjoyed your description of the Fractal Fantasies. However, I point out that some of what you said is almost impossible! I've calculated the number of rectangles and "spots" for every size board, using the formula: $(2N + 1) - 1$, where N is the degree of the board. From this, one can find the number of spots by simply multiplying by 6. Once you have this number, divide by two and subtract two to find the number of stones for each player. You have stated that grand masters have been known to use boards of degree 20. I've checked my calculations repeatedly, and this

would require each player start with 6,291,451 stones! Assuming each opening move (only those to place your pieces) took 2 seconds, the players wouldn't be able to move until 291.2708797 days after they started the game. Am I missing something, or are your numbers as ludicrous as they seem to me? Please don't take offense at this. I still found the concept very enjoyable.

Four Games for Gardner

Elwyn Berlekamp

Scientific American readers first became aware of the game of Domineering in [Gar74]. In the same article, Martin Gardner also presented another game called Quadraphag, which combined some features of chess with some features of Go. The present paper extends both of those themes (and others).

A 19 × 19 Go board may also be regarded as an 18 × 18 array of squares. If we cover over the middle two rows and the middle two columns, we then have four 8 × 8 arrays of squares. One of them is treated as a 9 × 9 Go board; the others are treated as 8 × 8 boards on which we will play Chess, Checkers, and Domineering, respectively. In Domineering, White (also known as Right) has legal moves consisting of placing a 2 × 1 domino onto any pair of horizontally adjacent empty squares of the board. His opponent is Black (also known as Left). Her legal moves consist of placing a 1 × 2 domino onto any pair of vertically adjacent empty squares. Empty squares are shown in white.

We assume the reader has some acquaintance with Checkers, Chess, and Go.

The problem to which this paper is devoted appears appears in Figure 1.

To start, the reader might wish to warm up by considering the four boards as four isolated, independent problems, one in each of the four games. However, our primary interest is in playing all four games at once. At each turn, the player may play any legal move on whichever board he chooses. A player who has no legal move must resign, and that ends the game.

Elwyn Berlekamp is an academic and entrepreneur, a theoretician and engineer, and has worked and played with games for over 50 years.

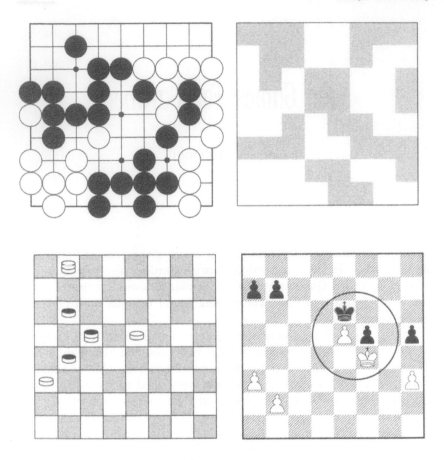

White to move

Figure 1. Initial position (in the Domineering board, the shaded areas are out of play).

This last-move-wins rule is standard in Checkers and in Domineering. It is also the governing criteria in the chess problem shown, because the kings and pawns shown within the circle in the middle of the board are deadlocked in a configuration that chess players call *mutual zugzwang*. So, for our purposes, we may treat only the simplified "mortal" chess problem, in which the encircled kings and pawns are removed from the board, the concept of stalemate is eradicated, and the goal (as in checkers) is to get the last legal move.

Although Go players are accustomed to passing and then counting score, the American Go Association rules require that a player must pay a one-point fee for each pass. This fee is conventionally paid by returning one of your opponent's stones that you have captured to the pot. After both players have made such passes, they might very well elect to count score by alternately filling in their territories to see who runs out of moves first. Thus, playing Go with the last-move-wins rule has no significant effect on the players' strategies. It is basically the same game, except for a minor procedural difference in how the score-counting is conducted. This difference often has no affect on the game's outcome.

A "move" in Go or Domineering corresponds to what is called a "ply" in chess or checkers. We use the Go terminology: One player plays only even-numbered moves; the other player plays only odd-numbered moves.

Rules buffs will no doubt wish to consider many subtleties, including the following:

1. North American and New Zealand Go have a "superko" rule, which bans any move that leads to any position that has already appeared earlier in the same game. Chinese Go also has a superko rule that bans some (but not all) such moves. The Japanese ko rule bans only moves that lead immediately back to the prior position in a 2-move loop. Japanese Go has no superko rule to ban 4-move loops, 6-move loops, etc. Professional games affected by these discrepancies are very rare. Since draws by repetition through cycles of 4 or more positions are fundamental to the strategy of checkers, we follow the Japanese ko rule, which legitimizes the possibility that the game might hang in a long loop (one with more than 2 moves). Since there is no possibility of a 2-move (i.e., 2-ply) loop in chess or in checkers, both games are compatible with the ko rule in Go.

2. What is the geographical scope of the ko-ban rule, which prohibits the immediate recapture of ko? A global interpretation allows recapture after an intermediate play on another board; a local interpretation leaves the ko-ban in force until there has been another play on the Go board.

3. What is the geographical scope of the "compulsory jump" rule in checkers? This rule states that if any jump move is possible, then nonjump checker moves are forbidden. According to a local interpretation, only nonjump checker moves are forbidden, and a

player might elect to play on another board even if a checkers' jump is possible. According to a global interpretation, a checkers' jump must take precedence over all other moves, on all boards.

Although the different variations of the rules entail major philosophical differences, it turns out that the vast majority of game positions are susceptible to analyses that do not depend on the details of the rules.

When the four problems are considered in isolation, we claim that White should draw Checkers, and win at each of Go, Chess, and Domineering. When all four boards are treated as a single game played together, we claim that White can win, no matter how one chooses to interpret rules 1, 2, and 3.

Details of these solutions will appear in [Now01].

Acknowledgements

I am indebted to Bill Spight and Gin Hor Chan for help in composing and debugging these problems, and to Silvio Levy for assistance in the preparation of this paper.

Bibliography

[BCG82] E. R. Berlekamp, J. H. Conway, and R. K. Guy. *Winning Ways for Your Mathematical Plays*, two volumes. Academic Press, London, 1982. Translated into German: *Gewinnen, Strategien für Mathematische Spiele* by G. Seiffert, Foreword by K. Jacobs, M. Reményi and Seiffert, Friedr. Vieweg & Sohn, Braunschweig, four volumes, 1985. Second edition, Natick, Mass.: A K Peters, 2000–2002, four volumes.

[Gar74] Martin Gardner. Mathematical games: Cram, crosscram and quadraphage: new games having elusive winning strategies. *Scientific American*, 230(2):106–108, February 1974.

[Now96] Richard Nowakowski, editor. *Games of No Chance: Combinatorial Games at MSRI, 1994*. MSRI Publications, 29, Cambridge University Press, 1996.

[Now01] Richard Nowakowski, editor. *More Games of No Chance*. MSRI Publications, Cambridge University Press, 2001. Sequel to [Now96], to appear.

The Sol LeWitt Puzzle: A Problem in 16 Squares

Barry Cipra

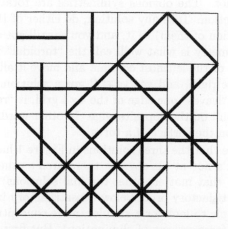

The contemporary conceptual artist Sol LeWitt often works with geometric designs. One of his etchings, *Straight Lines in Four Directions and All Their Possible Combinations* (1973), suggests a mathematical problem. Notice that some of the dark lines in the figure above (adapted from LeWitt's design) continue from one square to the next, but others don't. Is it possible to rearrange the 16 squares, keeping them in a 4×4 grid and not rotating any of them, so that all the lines go all the way from one edge of the grid to the other? If so, how many different solutions are possible? And if not, why not?

This question was first posed by the author in [Cip99, p. 113]. The puzzle can indeed be solved. There are three "geometrically distinct" solutions shown in Figure 1.

Barry Cipra is a freelance mathematics writer based in Northfield, Minnesota.

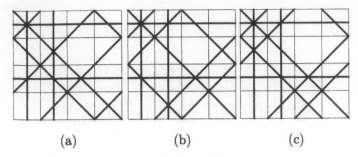

(a)　　　　　　　　　(b)　　　　　　　　　(c)

Figure 1.　　The three geometrically distinct solutions to the Sol LeWitt Puzzle.

We need to clarify what "geometrically distinct" means. There are several symmetries at work in the Sol LeWitt puzzle. Two of them are obvious, one is not. The obvious symmetries are rotation through 90 degrees and reflection: Take any solution, do either of these operations (or any combination of them) to it, and you've still got a solution. The non-obvious symmetry is what we'll call the "toroidal" property: Take the topmost row, or the leftmost column, and move it all the way to the bottom, or to the right, and you've still got a solution. In particular, when a diagonal "leaves" one edge of the 4×4 grid, it "re-enters" at the corresponding point on the opposite edge. In other words, each solution could be drawn on the surface of a torus.

It's easy to see that the three solutions in Figure 1 have this toroidal property. But it's not at all obvious there aren't other, non-toroidal solutions (or, for that matter, other toroidal solutions). We'll present a somewhat unsatisfactory proof below (unsatisfactory in that we'd like to encourage people to look for a simpler, more conceptual proof—ours is partly a brute-force process of elimination). But first, we need to introduce another, *non-geometric* symmetry that's hidden in the problem, which we'll call *existential* [Sar56] symmetry.

Notice that in each square, there either *is* a horizontal line or there *isn't,* and likewise for the vertical and two diagonals. What if we switch these? A little thought reveals that if you take a solution and erase all the horizontal lines that *are* in it and simultaneously draw horizontal lines in the squares where they *weren't,* then you've still got a solution! The same, of course, holds for the vertical and two diagonals.

If you like, there are three symmetry groups lurking within the LeWitt puzzle: the rotation/reflection group of order 8, a toroidal group of order 16, and an "existential" group of order 16. The first group is the most obvious. The third, once you see it, is also obvious. Only the second still requires proof.

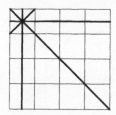

Figure 2.

So here's a proof that all solutions must have the toroidal property.

To begin with, the only thing at stake is the "toroidality" of the diagonals. But by reflection, it suffices to show this only for the "down" diagonals: If there were a solution with non-toroidal "up" diagonals, then by reflection through, say a vertical line, there'd be a solution with non-toroidal "down" diagonals. Finally, by the existential symmetries, we may assume the "full" square—the one with all four lines drawn in it—is in the upper left-hand corner. (Once we've proven toroidality, we've got a geometric way of moving the full square wherever we like, but for now all we've got is existentialism!) If we start by drawing the continuations of the lines in the full square, the picture looks like Figure 2.

Now where can the other four down diagonals be drawn? Up to reflection across the main diagonal (joining upper left to lower right), there are only four possibilities (Figure 3).

Note that the first two are toroidal. It suffices, therefore, to show why the other two cannot occur. Let's take them one at a time.

In Figure 3(c), the horizontal line in the top row goes through three squares with down diagonals. That means that the other horizontal line must be in the third row, since the other two rows each have two squares with down diagonals in them. A similar argument shows that the other vertical line must be drawn in the third column, since only it has two squares with down diagonals in it. The result is as in Figure 4(a).

But this has three squares (the shaded ones) with the combination of vertical, horizontal, and down diagonal, and that can't be.

Similarly, in Figure 3(d), the second horizontal and vertical lines are forced to be in the fourth row and column, respectively, resulting in the same contradiction (Figure 4(b)).

From this we conclude that the down diagonals must have the toroidal property, and that concludes the proof.

Now why are there only three geometrically distinct solutions? (It may not be immediately obvious that the three solutions in Figure 1

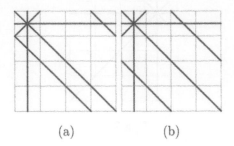

(a)   (b)

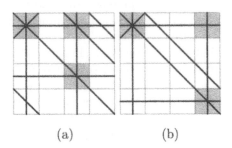

(c)   (d)

Figure 3.

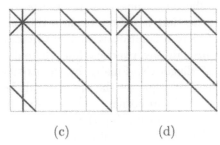

(a)   (b)

Figure 4.

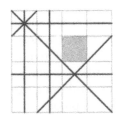

Figure 5.

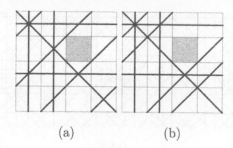

(a)   (b)

Figure 6.

really are geometrically distinct, but we leave it to the reader to convince himself they are.) Our proof relies on toroidality.

We begin with a little lemma: The four squares with both horizontal and vertical lines drawn in cannot be positioned at the corners of a square of any size. We leave it to the reader to convince himself that it's impossible to draw the up and down diagonals in such an arrangement without duplicating one of the combinations (and omitting another).

The upshot of this is that the two lines in one direction, say the verticals, must be toroidally adjacent, and the other two toroidally non-adjacent. Using either toroidal or existential symmetry, we can assume the full square is in the upper left-hand corner, so the picture starts out like Figure 5.

Note that we've shaded in one of the two candidates for the blank square. This is because the two are actually equivalent under our geometric symmetries: To interchange them, do a reflection across a horizontal line, and then move the row with the full square back to the top. Consequently, we may assume the shaded square is the one to leave empty. This leaves only two possibilities for the other up diagonal, as shown in Figure 6.

In both, we have two squares with just a horizontal line, so we need to put the final down diagonal through one of them. But in Figure 6(a), it can't go through the shaded square (because that's the one we've chosen to leave empty), so there's only one choice, which produces the solution in Figure 1(a).

In Figure 6(b), on the other hand, either choice is OK. Hence we get the other two solutions in Figure 1(b,c).

Because of toroidality, each of our three solution can be used to tile the plane (Figures 7(a,b,c)).

(Note: Figures 7(a) and 7(b) look a lot alike—they may appear to be related by a reflection—but they really are geometrically distinct. We'll see below two reasons why they appear so similar.)

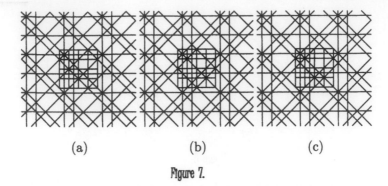

(a)   (b)   (c)

Figure 7.

Finally, if we let H, V, U, and D represent the existential symmetries for the horizontal and vertical lines and the up and down diagonals, respectively, it's easy to show that, applying these to our three geometrically distinct solutions of Figure 1(a–c), we have $H(a) = H(b)$, but all other actions are trivial. (The equality must be interpreted as allowing for rearrangement by the geometric symmetries.) In particular, even if you allow for all possible symmetries to act, there are still two essentially different solutions: the pair (a)/(b) and the singleton (c). This helps explain why Figures 7(a) and7(b) look so similar: moving the horizontal lines of one turns it into (a mirror image of) the other.

The three geometrically distinct solutions were first identified by John Conway of Princeton University. (The proof given here that there are no others is, in effect, a streamlined version of Conway's argument.) Jonathan Needleman, a student at Oberlin College, has pointed out one additional peculiar property of the three solutions: If you move just one vertical line, say from the second to the fourth column (which keeps the vertical lines toroidally adjacent), Figure 1(a) again becomes 1(b), while Figure 1(c) again remains fixed. This gives another explanation for the deceptive similarity of the tilings in Figures 7(a) and 7(b).

A priori, it's not clear that Needleman's operation (moving one line) should produce a solution at all. It would be nice to have a direct, conceptual proof of this, along with the one mentioned earlier, for the toroidal property.

We close by posing two addition problems similar to the LeWitt puzzle. The first was dreamt up by the author. The second was posed by Loren Larson, a mathematician and problems expert at St. Olaf College in Northfield, Minnesota.

Figure 8(a) consists of the 16 combinations of quarter circles either drawn or not drawn at the four corners of a square. Is it possible to rearrange the 16 squares, again without rotating any of them, to pro-

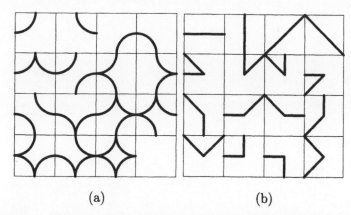

(a)   (b)

Figure 8.   Two more puzzles, the second proposed by Loren Larson.

duce a pattern of complete circles and semi-circles—preferably with the toroidal property, so that each semi-circle on one edge matches up with one on the opposite edge?

In Figure 8(b), the pattern in each of the 16 squares consists of two straight lines emanating from the center of the square to two of eight points on the sides (the four corners and the four midpoints). The patterns in the top row are invariant under rotation by 180 degrees. The other patterns are not. The problem here is to rearrange the squares, allowing for 180-degree rotations (but not rotations of 90 degrees), to produce one continuous, non-self-intersecting dark line, with turns only at the centers and the corners of the squares.

Bibliography

[Cip99] Barry Cipra. *What's Happening in the Mathematical Sciences 1998–1999*, volume 4. American Mathematical Society, 1999.

[Sar56] Jean Paul Sartre. *Being and Nothingness*. Philosophical Library, 1956.

Sum-Free Games

Frank Harary

Dedicated to my games mentor, Martin Gardner,
on his 85th birthday.

A *sum-free set* of positive numbers never has two distinct numbers which add up to a third in the set. So, $\{1, 3, 5, 7, \ldots\}$ is sum-free (since adding two odd numbers always yields an even number), while $\{1, 3, 5, 6\}$ is not sum-free (since $1 + 5 = 6$).

In these games, two players, Alice and Bob, alternately place numbers in numeric order from $\{1, 2, 3 \ldots, 8\}$ into one of two columns, C_1 or C_2. They try to maintain the sum-free status by keeping the three number combinations from occurring in one column. In this game, a player loses if, after moving, two numbers add to a third in one column.

The Achievement Game

In this game, Alice moves first by placing number 1 in one of the two columns C_1 or C_2. This move has absolutely no effect on the outcome of the game, as the names of the two columns do not matter, making it what I call (in jest) a "shrewd move," We'll assume Alice placed 1 in column C_1.

Now, Bob can place 2 in whichever column he pleases. If he is wise, he will choose to place 2 also in C_1, for this moves leads to victory. The

---

Frank Harary applied graph theory to anthropology, chemistry, and computing, and discovered that some theorems can be played as two-player Achievement and Avoidance games.

rule of the game is that **no move is permitted that would destroy the sum-free status of either column.**

Hence, with both 1 and 2 in C_1, Alice must place 3 in column C_2. This forced move is denoted 3*.

In the *Achievement Game,* the last player who can move wins. In the *Avoidance Game,* the last player who can move loses.

We'll prove that Bob can win the Achievement Game no matter what moves Alice makes. Let's see the possibilities. We'll show in table form the opening moves described so far:

C_1	C_2
1	3*
2	

Were Bob to place 4 in column C_1, he'll lose on move 8 (try it out!) But a wise Bob will place 4 in column C_2, and afterward Alice can place 5 in either column. In either case, Bob should put 6 in the other column: In the first case, his move is forced (since $1 + 5 = 6$ he can't place 6 in the first column), in the other case, it merely shows forethought:

1	3*		1	3*
2	4		2	4
5	6*		6	5

In either case, Alice cannot place 7, so Bob wins the Achievement Game.

The Avoidance Game

This time, the player who makes the last move loses!

We shall prove that Alice can always win no matter what move Bob makes. The proof is a bit more complex that that for the Achievement game. If Alice starts in C_1, Bob can play his response in either column.

We'll consider first, C_1. Then Alice must put 3* in C_2, and Bob now has two choices for 4. If he places 4 in C_1, 5* and 6* are forced, and Alice *should* play 7 in column 2 forcing Bob to play the last move in column 1:

$$
\begin{array}{cc}
1 & 3^* \\
2 & 5^* \\
4 & 6^* \\
8^* & 7
\end{array}
$$

If, on the other hand, Bob played 4 in the second column, Alice put 5 in C_1, forcing Bob's 6^* C_2. As Alice cannot move, she wins:

$$
\begin{array}{cc}
1 & 3^* \\
2 & 4 \\
5 & 6^*
\end{array}
$$

So, if Bob plays his second move in C_1, Alice can guarantee a win. Suppose Bob plays his second move in C_2. Then Alice wins more simply. Alice puts 3 in C_1. Bob is forced to move 4^* in C_2. Then Alice puts 5 in C_2 and Bob must write 6^* in C_1:

$$
\begin{array}{cc}
1 & 2 \\
3 & 4^* \\
6^* & 5
\end{array}
$$

And since Alice cannot place 7 in either column, she wins.

Variations

There are several variations to the Achievement and Avoidance Games. Naturally, these variations can be combined to make more variations than appear here.

V1 (Variation 1) Start with another set of consecutive integers, in place of $\{1, 2, \ldots, 8\}$. For instance, $\{1, 2, \ldots, 10\}$ or $\{3, 4, \ldots 12\}$.

V2 We could remove the word "distinct" in the definition of sum-free. Under the old definition, the set $\{1, 2\}$ is sum free, but under this new one it is not, since $1 + 1 = 2$.

V3 The number c of columns can be 3 or 4 or even more. (More than 3 columns apparently give games that are too long to remain interesting.)

V4 These can be played with 3 players. It seems that games with more players would be exceedingly boring.

V5 In *choice games,* the players can choose any remaining number to play, and need not play the numbers in order. So the first move might be to place 4 in the first column. The next player can place any number *except* 4.

Unsolved Problems

- Who wins under variations of the Avoidance and Achievement games under variations under different choices of variations **V1** and **V3**? By a result from game theory, there is always a winner. For games with three players, when can one player force a win? When is one player forced to lose?

- For each game, what is the corresponding Ramsey-type number? How large a starting set do you need before it is guaranteed that a non-sum-free situation will occur (no matter how unskillful the players).

Bibliography

[GHR] G. Gupta, F. Harary, and D. Ranjan. Computation of sum-free sets of numbers. (to appear).

[GRS90] R. L. Graham, B. L. Rothschild, and J. H. Spencer. *Ramsey Theory.* Wiley, New York, 2nd edition, 1990.

Shadows and Plugs

Gwen Roberts

Martin Gardner has powerfully influenced my teaching of high school mathematics. Perhaps the most important is in the joy I see as students learn math using puzzles and games inspired by him. May the joy Martin experienced as a self-described "amateur" Mathematician continue to filter down to generations of students who have trouble learning by the textbook alone.

My interest in mathematics began in 1958 at Tulsa Central High School in Oklahoma when I studied Plane and Solid Geometry. I was excited to imagine how infinite space can be made up of points, entities with no dimension. Amazing!

By doing constructions in three-space, students were really learning about mathematical rigor and the nature of mathematical proof.

This paid off later when we tried to visualize the three-dimensional solid resulting from revolving a line or semicircle about the x-axis which we used for integration in Calculus. Though we had memorized volume formulas for cylinders, cones, and spheres, we had no idea why the fractions 1/3 or 4/3 were involved until we learned about integration.

I married and raised three sons, and for the following thirty years my main connection to mathematics was Martin Gardner's (a former Tulsan!) column in *Scientific American*. I played Sprouts. I enjoyed Conway's Game of Life. The Mathematical Games column was my introduction to M. C. Escher. Marjorie Rice (an amateur, like me, who worked with tessellations) became one of my heroes. Martin Gardner connected me to a world of puzzles, games, mathematics, art, magic, and science. He was my link to a world of exciting ideas.

Gwen Roberts is a high school math teacher. She creates puzzles with Robin Lamkie.

399

In 1988, I began a teaching career. Sadly, I found that Solid Geometry was no longer part of the high school curriculum. I attempted to make up for this absence by using manipulatives to help students grasp abstract concepts. A kinesthetic learning takes place when students play with physical representations. Public school budgets are limited and commercially available models are expensive, so I found that I needed to make manipulatives for my students. In addition, some concepts had no ready-made physical models.

My calculus students struggled with visualizing disks, washers, and cylindrical shells. They struggled because they had not had the advantage and experience of visualizing loci. I made my Plugs and Shadows puzzles to help them:

I show them the three holes, and ask the students to visualize an object that snugly fits through all three.

I made many dissection puzzles to help students in my geometry classes see that a linear change (say in the side of a polygon) causes the area to change in a dramatic but predictable way. The first (and most popular) physical model I made was suggested by Martin Gardner's discussion of Dudeney's four piece dissection of equilateral triangle to square [Gar61, Ch. 3]:

This puzzle has many applications. A good problem is, "Given a square and equilateral triangle with equal area, how do the length of the sides of each figure relate?"

For years I've taught a course called Transformational Geometry. Students must be able to see in their mind's eye the changes involved in reflections, rotations, and size transformations. Once these images are clear in a student's mind, concepts from group theory seem to come more easily. The Project Mathematics Video series by Tom Apostol and Jim Blinn uses animation to bring these changes to life. One of my students became fascinated with the Hinge Proof of the Pythagorean theorem from the videos. We assembled a physical model using duct tape for the hinges. I've used this puzzle for years as an example of a compelling visual proof (as opposed to the two-column proofs given in most high school geometry texts):

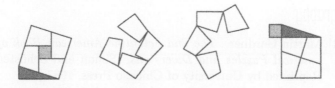

Another student's observation of a triangle dissection involving a surprising conclusion involving the 1:7 ratio was added to my growing collection of demonstration puzzles:

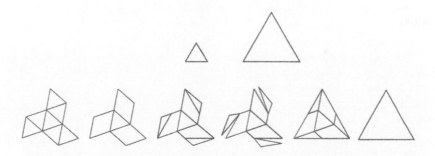

A basic task in Calculus is calculating the volume of a solid of revolution obtained by rotating any function $f(x)$ about the x-axis. The function could describe a triangle, a rectangle, a semicircle, a cross, a letter of the alphabet, or even the profile of a face. To calculate the volumes, students must be able to visualize stacking narrow discs to build the solids.

In the "triangle, circles, and square puzzle" shown below, the students discover a surprising property of the relationships of the related volumes of revolution. The volumes of the cone, sphere, and cylinder are related in the ratio of 1 : 2 : 3.

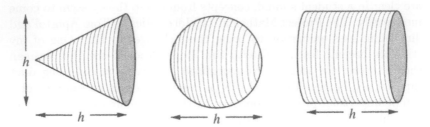

Bibliography

[Gar61] Martin Gardner. *The 2nd Scientific American Book of Mathematical Puzzles and Diversions.* Simon and Schuster, 1961. Reprinted by University of Chicago Press, 1987.

A Neglected Trigonometric Gem

Eli Maor

In memory of Jon Froemke (1941–1998), colleague, puzzlist, and trusted friend.

> In his very numerous memoirs, and especially in his great work, *Introductio in analysin infinitorum* (1748), Euler displayed the most wonderful skill in obtaining a rich harvest of results of great interest... —E. W. Hobson, *"Squaring the Circle": A History of the Problem* (1913)

Among the numerous infinite series to come out of Euler's creative mind, there is the following little-known identity:

$$\frac{1}{x} - \cot x = \frac{1}{2}\tan\frac{x}{2} + \frac{1}{4}\tan\frac{x}{4} + \frac{1}{8}\tan\frac{x}{8} + \cdots \tag{1}$$

To prove it, we begin with the double-angle formula for the cotangent,

$$\cot 2x = \frac{1 - \tan^2 x}{2\tan x} = \frac{\cot x - \tan x}{2}$$

Starting with an arbitrary $x \neq n\pi/2$ and applying the formula repeatedly, we get

$$
\begin{aligned}
\cot x \;=\;& \tfrac{1}{2}\left(\cot\tfrac{x}{2} - \tan\tfrac{x}{2}\right) \\
=\;& \tfrac{1}{4}\left(\cot\tfrac{x}{4} - \tan\tfrac{x}{4}\right) - \tfrac{1}{2}\tan\tfrac{x}{2} \\
=\;& \tfrac{1}{8}\left(\cot\tfrac{x}{8} - \tan\tfrac{x}{8}\right) - \tfrac{1}{4}\tan\tfrac{x}{4} - \tfrac{1}{2}\tan\tfrac{x}{2} \\
=\;& \cdots \\
=\;& \tfrac{1}{2^n}\left(\cot\tfrac{x}{2^n} - \tan\tfrac{x}{2^n}\right) - \tfrac{1}{2^{n-1}}\tan\tfrac{x}{2^{n-1}} - \cdots - \tfrac{1}{2}\tan\tfrac{x}{2}
\end{aligned}
$$

Eli Maor teaches mathematics and the history of mathematics at Loyola University in Chicago.

As $n \longrightarrow \infty$, $\frac{1}{2^n} \cot \frac{x}{2^n}$ tends to $1/x$, so we get

$$\cot x = \frac{1}{x} - \sum_{n=1}^{\infty} \frac{1}{2^n} \tan \frac{x}{2^n}$$

from which Equation 1 follows.

If Equation 1 is not particularly remarkable, an immediate consequence of it certainly is: If we substitute $x = \frac{\pi}{4}$, we get

$$\frac{4}{\pi} - 1 = \frac{1}{2} \tan \frac{\pi}{8} + \frac{1}{4} \tan \frac{\pi}{16} + \frac{1}{8} \tan \frac{\pi}{32} + \cdots$$

Replacing the 1 on the left side with $\tan \frac{\pi}{4}$, moving all the tangent terms to the right side and dividing the equation by 4, we get

$$\frac{1}{\pi} = \frac{1}{4} \tan \frac{\pi}{4} + \frac{1}{8} \tan \frac{\pi}{8} + \frac{1}{16} \tan \frac{\pi}{16} + \cdots \qquad (2)$$

Equation 2 must surely rank among the most beautiful in all of mathematics, yet it does not appear in trigonometry or calculus textbooks; nor is it listed in L. B. W. Jolley's *Summation of Series* [Jol61], a compilation of over a thousand arithmetic, algebraic, and trigonometric series. What is more, the series on the right side converges extremely rapidly (note that the coefficients **and** the angles decrease by a factor of 2 with each term), so we can use Equation 2 as an efficient means to compute π: it takes just twelve terms to obtain π correct to six decimal places, that is, to one millionth; four more terms will increase the accuracy to one billionth.[1]

If we remultiply Equation 2 by 4 and introduce $p = \pi/4$, we get the even simpler-looking equation

$$\frac{1}{p} = \tan p + \frac{1}{2} \tan \frac{p}{2} + \frac{1}{4} \tan \frac{p}{4} + \cdots \qquad (3)$$

Equation 3 is reminiscent of the famous Runner's Paradox, proposed by the philosopher Zeno of Elea in the third century B.C.: A runner attempts to cover a given distance (in this case, of length 2) by first covering one half this distance, then half of what remains, and so on ad infinitum. The total distance covered is the sum of these partial segments, that is, the sum of the infinite progression $1 + \frac{1}{2} + \frac{1}{4} + \frac{1}{8} + \cdots$. The Greeks knew that the sum of this progression is 2, but they could not explain why an infinite sum can have a finite value. Zeno's paradoxes were meant to point out the inability of his contemporaries to deal with infinite processes.

---

[1]One may raise the objection that Equation 2 expresses π in terms of itself. However, since the trigonometric functions are immune to the choice of angular units, we can use degrees instead of radians and write Equation 2 in the form $\frac{1}{\pi} = \frac{1}{4} \tan 45° + \frac{1}{8} \tan 45/2° + \frac{1}{16} \tan 45/4° + \cdots$

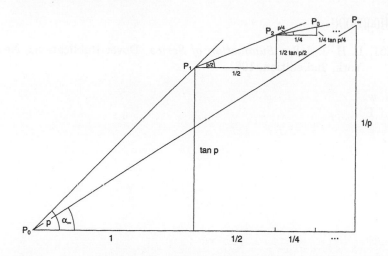

Figure 1. A geometric interpretation similar to the Runner's Paradox.

We can give Equation 3 a geometric interpretation similar to the Runner's Paradox (Figure 1): The runner starts at point P_0, goes a distance of one unit to the right and one unit up, arriving at P_1. The vertical distance covered is $1 = \tan p$. From point P_1, the runner goes a distance $\frac{1}{2}$ to the right followed by $\frac{1}{2}\tan\frac{p}{2}$ up, arriving at point P_2. Continuing in this manner, we get a staircase route with the horizontal steps and angles shown decreasing geometrically by a factor of 2 with each step. Eventually, the runner will arrive at a limiting point P_∞, whose height above the starting point is the sum of the series of Equation 3.

We might ask, what is the straight-line direction from the initial point to the final point? Denoting the angle between the horizontal line and the line segment P_0P_∞ by α_∞, we have

$$\tan\alpha_\infty = \frac{\tan p + \frac{1}{2}\tan\frac{p}{2} + \frac{1}{4}\tan\frac{p}{4} + \cdots}{1 + \frac{1}{2} + \frac{1}{4} + \cdots}, \tag{4}$$

so we can interpret $\tan\alpha_\infty$ as a weighted average of the tangents appearing in the numerator. These tangents are the slopes of the line segments P_0P_i (not shown in the figure). In view of Equation 3, we can replace the numerator by $\frac{1}{p}$, getting

$$\tan\alpha_\infty = \frac{\frac{1}{p}}{1 + \frac{1}{2} + \frac{1}{4} + \cdots} = \frac{1}{2p} = \frac{2}{\pi} \approx .637,$$

or $\alpha_\infty \approx 32.48°$.

Bibliography

[Jol61] L. B. J. Jolly. *Summation of Series*. Dover Publications, New York, 2nd edition, 1961.

Kotani's Ant Problem

Dick Hess

A recent postcard from a friend, Nob Yoshigahara in Tokyo, asked me to imagine an ant crawling on the walls, floor and ceiling of a $1 \times 1 \times 2$ rectangular room as shown in Figure 1. An ant is positioned at A and can travel to any other point on the surface of the room. I was asked to find the farthest point from A and was told it is not the point B! This problem is the creation of Yoshiyuki Kotani, a mathematics professor in Saitama, Japan, and he discovered that the farthest point is not at B but at a specific place on the far face.

Recreational mathematics problems involving geodesics on the walls of rectangular rooms have been around for nearly 100 years and are well documented by Singmaster [Sin93]. They are often referred to as spider and fly problems.

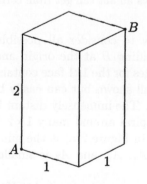

Figure 1.   A $1 \times 1 \times 2$ room.

Dick Hess is a long-time enthusiast of recreational mathematics and mechanical puzzles.

As a warmup, let's work out the following problem:

Problem 1 *In the $1 \times 1 \times 2$ rectangular room, what is the shortest path between diagonally opposite corners A and B?*

This problem can be analyzed conveniently by unfolding the room along edges in various ways as shown in Figure 2. Geodesic paths in the room are seen simply as straight lines on the unfolded room walls. Care is required, however, to consider all possible ways to unfold the room to determine the true shortest path between two points. For example, in Figure 2(a) the path between A and B has a length of $\sqrt{2^2 + 2^2} \approx 2.8284$ while in Figure 2(b) it has length $\sqrt{1^2 + 3^2} \approx 3.1623$. In the case of the second path, the ant hasn't taken the shortest geodesic route.

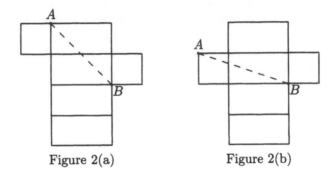

Figure 2(a)   Figure 2(b)

Figure 2.   Two ways an ant can get from corner A to corner B.

Figure 3 shows how to consider all possible paths from A to B. It unfolds the room holding B at the origin and indicates the infinite number of possible images for the 1×1 face containing A. In this diagram the 1×2 sides are not all shown but can easily be imagined in place to plot paths from A to B. The immensely distant images of A correspond to paths from B that spiral around many 1×2 faces. From the Figure it's clear that the path in Figure 2(a) is the minimum; the distance $\sqrt{8}$ from B to the images A_1, A_4, and A_7 is smaller than the distance to any other image of A.

This same approach can now be applied to Kotani's original problem:

Problem 2 *Where on the $1 \times 1 \times 2$ room is the farthest from corner A and what is the distance between it and A? (Answer on page 411.)*

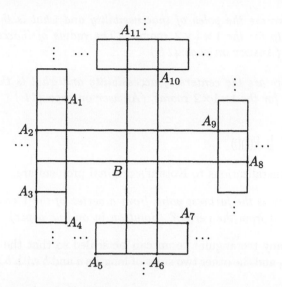

Figure 3. The infinite number of ways to get from B to A along the walls, floor and ceiling of a $1 \times 1 \times 2$ room.

Related problems of interest on the $1 \times 1 \times 2$ room are:

Problem 3 *What point is farthest from the center point of a 1×1 face and how far away is it? (Answer on page 411.)*

Problem 4 *What point is farthest from the center point of a 1×2 face and how far away is it? (Answer on page 411.)*

Poles of Inaccessibility and Center of Accessibility

It is interesting to define two types of special points for the room. First, there is a pair of points in the room that are more distant from each other than any other pair of points. These points are called *poles of inaccessibility* and the distance between them is called the *radius of inaccessibility*. If the ant and his girlfriend are located at the poles of inaccessibility they are as far apart as possible in the room. Second, there are points on the room's surface having the smallest distance to their farthest points. These points are the centers of accessibility. If the ant is at a *center of accessibility* his girlfriend can be placed no farther from him than the *radius of accessibility*.

Problem 5 *Where are the poles of inaccessibility and what is the radius of inaccessibility for the $1 \times 1 \times 2$ room? (The radius of inaccessibility exceeds 3.01.) (Answer on page 411.)*

Problem 6 *Where are the centers of accessibility and what is the radius of accessibility for the $1 \times 1 \times 2$ room? (Answer on page 411.)*

The $1 \times a \times b$ Room

The natural generalizations to Kotani's original problem are,

Problem 7 *Where is the farthest point from a vertex of the $1 \times a \times b$ room and how far is it from the vertex? (Solution in on-line paper)*

Naturally, any rectangular room can be scaled so that the shortest side length is 1, and the other two sides of lengths a and b with $b \geq a \geq 1$.

Problem 8 *For each face, what point is the farthest from the center point of the face and how far away is it? (Solution in on-line paper)*

Problem 9 *For which rooms is the farthest point from a face center the center point of the opposite face? (Solution in on-line paper)*

Problem 10 *For which room is the farthest point from a face center nearest to an edge? (Solution in on-line paper)*

Other Problems to Consider

Problem 11 *Where are the poles of inaccessibility and what is the radius of inaccessibility for a $1 \times a \times b$ room?*

The answer to this problem is known subject to the following unproven conjecture (Solution in on-line paper):

Conjecture 1 *The two points of a pole pair always reside on opposite $1 \times a$ faces and are mirror points of each other.*

A computer program written to search numerically for pole pairs was run for a large variety of rooms and produced results always obeying this conjecture. No mathematical proof of this conjecture is available, but its truth isn't too surprising; a proof or counterexample would be greatly appreciated.

Problem 12 *For which rooms are the diagonally opposite corners poles of inaccessibility? (The answer is known only if one assumes the conjecture.)*

Problem 13 *Where are the centers of accessibility and what is the radius of accessibility for a $1 \times a \times b$ room? (This is an open question.)*

Solutions

These terse answers are provided so you can check them against your own answers. For complete solutions, refer to the on-line site for the book: http://www.g4g4.com

Answer 1 *(Problem 1 on page 408)* The solution is given in the text following the problem.

Answer 2 *(Problem 2 on page 408)* $\sqrt{130}/4$

Answer 3 *(Problem 3 on page 409)* The farthest point is at the center of the far face, 3 units away.

Answer 4 *(Problem 4 on page 409)* The farthest point is at $x = 1/6$, $y = 1/2$ on the far face, at a distance of $13/6$.

Answer 5 *(Problem 5 on page 410)* Assuming the poles are mirror points (not proved in general, but no counterexample is known), the radius of inaccessibility is $2 \cdot \sqrt{4 - \sqrt{3}}$.

Answer 6 *(Problem 6 on page 410)* The centers of accessibility are the center points of the four 1×2 faces. The radius of accessibility is $13/6$.

Bibliography

[Sin93] David Singmaster. *Sources in Recreational Mathematics, 6th edition*. Spider and Fly Problems, pp 176–177, November 1993. Self published. Updates will be available by CD from zingmaster@sbu.ac.uk.

Y2K Problem of
Dominoes and Tatami Carpeting

Yoshiyuki Kotani

Domino tiling is a little fun for those of us who love mathematics and puzzles, but perhaps it is also meaningful for Japanese people in their daily life because of the shape of tatami carpeting. The usual goal in Domino tiling problems is to fill a figure with dominoes, i.e., 1×2 rectangles. Combinatorists will often ask, "How many ways can the shape be tiled?" Here, we reverse the question and say, "Find shapes which have exactly 2000 tiling solutions."

The Problem

The goal is to tile any figure which consists of unit squares with 1×2 rectangles without overlapping or overhanging. When counting solutions, we do not unify equivalent solutions under symmetries. For example, the 2×3 rectangle has three solutions:

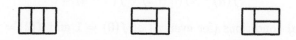

Yoshiyuki Kotani is a member of ARMJ (Academy of Recreational Mathematics, Japan) and CSA (Computer Shogi Association), and is Professor of Computer Science.

$2 \times n$, $3 \times n$, and $m \times n$ Rectangular Tiling

It is well known that the number of ways to tile a $2 \times n$ rectangle with dominoes forms the familiar Fibonacci sequence:

2×1	2×2	2×3	2×4	2×5	2×6	2×7	2×8	2×9	...
1	2	3	5	8	13	21	34	55	...

To see why, let $F(n)$ be the number of such tilings. If a vertical tile covers the two leftmost squares, a $2 \times (n-1)$ rectangle remains to be tiled. If two horizontal tiles cover the leftmost squares, a $2 \times n-2$ rectangle remains untiled. This leads us to the recursive formula:

$$F(n) = F(n-1) + F(n-2)$$

The number of tilings, call it $f(n)$, of $3 \times n$ rectangles can also be analyzed. Let $g(n)$ count the number of ways to tile a the shape formed by removing a single corner square from a $3 \times n$ rectangle. Focusing on how the top two squares at the leftmost end of a $3 \times n$ rectangle is covered (marked a and b below), we obtained the geometrical recurrences:

which translated into the recurrence formulae:

$$\begin{aligned} f(n) &= g(n-1) + f(n-2) + g(n-1) \\ g(n) &= f(n-1) + g(n-2) \end{aligned}$$

Removing $g(n)$, we get the recursion:

$$f(n) - 4f(n-2) + f(n-4) = 0$$

with initial conditions (for even n) of $f(0) = 1$ and $f(2) = 3$. We can then obtain:

3×2	3×4	3×6	3×8	...
3	11	41	153	...

(Naturally, when n is odd, the rectangle cannot be tiled.) This method can also be applied to obtain recurrences for $4 \times n$ rectangles, $5 \times n$, and so forth. But the analysis quickly gets quite complicated. But there is a great general formula which gives the number of $m \times n$ domino tilings which seems to have been known in physics since 1961 [Hos86].

Algorithms for Counting

We can define a simple function which counts how many ways there are to tile any figure f. The algorithm is based on recursive calls which calculate count two subshapes made by removing the vertical and horizontal dominoes that cover the upper-leftmost square, and adds these two counts.

Although the combinatorial explosion makes this algorithm extremely inefficient, the computer science technique of *memoizing,* helps to bring the combinatorial explosion down a bit. The idea is to record calculated values in a table so that they need not be recalculated when a shape reappears during another recursive call. A hash table is a good storage mechanism for these values. Note here, that it's perfectly all right if a hash table entry is overwritten by another; it simply means that if the first entry is needed again, it must get recalculated. Similarly, the same hash table can be used without reinitializing it. The entries may even come in handy when calculating similar shapes.

Y2K Problem

I posted this problem at a puzzle party as a competitive quiz event. The number 2000 is not so meaningful mathematically, but was certainly timely. The problem is:

> Find a shape which has exactly 2000 domino tilings. Use as few dominoes as possible.

Many people found the solution below, factoring 2000 into $2 \cdot 2 \cdot 2 \cdot 2 \cdot 5 \cdot 5 \cdot 5$. Some found the improvement using 19 dominoes by replacing three 2×2 squares by a 2×5 rectangle which has can also be tiled in 8 ways.

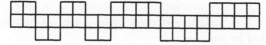

The problem was discussed in JARM (*Journal of the Academy of Recreational Mathematics*, Japan) and in NOBNET (puzzlers' mailing list). Many 18-domino solutions were reported, two examples of which are shown below.

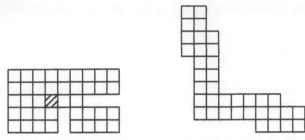

The one on the left was a solution which fits in a small rectangle, found by M. Odawara by hand. The second is a symmetric one found by W. H. Huang, who thought of using snake-like zigzags to generate various numbers. B. Harris showed a heuristic shape construction of a specific number of patterns by generating the Fibonacci sequence backward from the number keeping close to the golden ratio between adjacent pairs.

I know of two 17-domino solutions shown below.

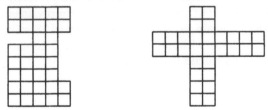

The one on the left is mine, obtained by programming a brute-force search of shapes which are made by removing six unit squares from a 5×8 rectangle. T. Arimatsu reported the second, cross-like shape. I think he found it by varying the lengths of the four arms.

Figures for Any Number of Tilings

Does there exist a figure for any number, N, of domino tilings? If so, how few dominoes are necessary to make it? B. Harris answered the first question by providing the sequence in Figure 1, where each shape consisting of N dominoes has exactly N ways to tile it. To see this, if the pair of unit squares marked "$\times$" is covered by a single domino, the remaining have $N - 1$ tiling patterns. If by a pair of dominoes, the rest can only be tiled in one way.

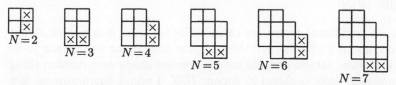

Figure 1. A sequence of patterns with exactly N domino tilings due to B. Harris.

Little is known about the second question. Figure 2 shows small shapes for $N \in \{2, \ldots 27\}$.

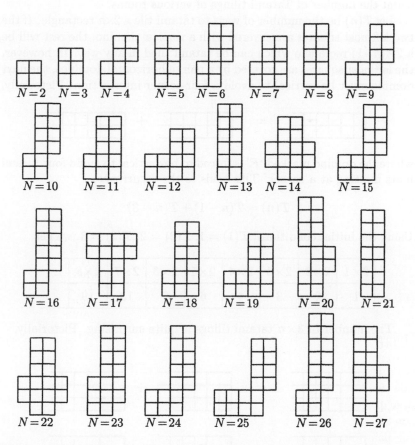

Figure 2. Figures close to the minimum number of squares which have exactly N domino tilings.

Tatami Tiling

Japanese traditional rooms are carpeted by tatamis. It is a thick carpet which is about 3×6 feet (90×180 cm), just enough for one person to lie on. Therefore, tatami carpets have a domino shape and domino tiling is a practical daily problem in Japan. (OK, I admit the argument is a little far-fetched....)

Although usual domino tilings are possible, there is an additional custom in smaller rooms to avoid criss-cross lines made by the edges of tatamis when four corners meet at a point. The reason for this custom is not well known, but we can observe it everywhere. Perhaps such corners can lead to mishaps when the corners twist or come apart. Let's try to count the number of Tatami tilings of various rooms.

Let $T(n)$ be the number of ways to tatami tile a $2 \times n$ rectangle. If the two leftmost squares are covered with a vertical domino, the rest will be a $2 \times (n-1)$ rectangle which can be tatami tiled in any way. If, however, the leftmost squares are covered by a pair of horizontal dominos, the next domino must be vertical to avoid a four-corner intersection. Pictorially,

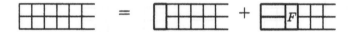

where the domino marked 'F' is forced to be vertical to avoid four tatami mats meeting at a corner. This leads to the recurrence,

$$T(n) = T(n-1) + T(n-3)$$

Using the initial conditions, $T(1) = 1$, $T(2) = 2$, $T(3) = 3$, we get

2×1	2×2	2×3	2×4	2×5	2×6	2×7	2×8	...
1	2	3	4	6	9	13	19	...

The number of $3 \times n$ tatami tilings is quite surprising. Pictorially,

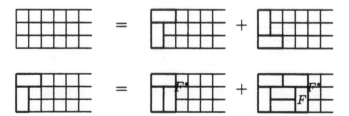

The domino marked F was forced by the tatami constraint. Although not yet placed, dominos will also be forced at the places marked F^*. Let $t(n)$ count the number of tatami tilings of a $3 \times n$ rectangle. Let $u(n)$ count the number of tilings of a $3 \times n$ rectangle with one corner square removed. The above pictures give the recurrences,

$$
\begin{aligned}
t(n) &= 2u(n-1) \\
u(n) &= \frac{1}{2} \cdot t(n-1) + \frac{1}{2} \cdot t(n-3)
\end{aligned}
$$

The $\frac{1}{2}$'s come from the forced tiles F^* halving the number of ways to tile the remaining rectangle. Plugging the second equation in for u in the first equation yields,

$$
t(n) = t(n-2) + t(n-4)
$$

and the Fibonacci recurrence emerges again! Using the initial conditions $t(2) = 2$, $t(4) = 4$, and the exceptional 3×2 rectangle which can be tiled with three horizontal dominoes, we get:

3×2	3×4	3×6	3×8	3×10	3×12	3×14	3×16	...
3	4	6	10	16	26	42	68	...

This is interesting from another point of view. The figures below show the real tatami carpeting preferred by Japanese people:

(A)

(B)

The $3 \times n$ rectangle tiling is just an arbitrary sequence of 3-tatami room tilings (A) and 6-tatami room tilings (B). They play the role of vertical dominoes and horizontal domino pairs in $2 \times n$ rectangle construction. For example, placing A, A, B, and A in order, The $3 \times 2n$

is constructed as the $2 \times n$

This means that, once mirror images are included, the number of $3 \times 2n$ tatami tilings is twice the number of $2 \times n$ domino tilings.

Open Problems

I leave you with two open problems:

1. Find a Y2K tatami tiling.

2. Give a formula for the number of tatami tilings for an arbitrary any $m \times n$ room.

Bibliography

[Gra81] Ron L. Graham. Fault free tilings of rectangles. In David Klarner, editor, *The Mathematical Gardner*, pages 120–126. Prindle, Weber, and Schmidt, Boston, 1981.

[Hos86] Haruo Hosoya. Matching and symmetry of graphs. symmetry: unifying human understanding. *Computers & Mathematics with Applications, Part B*, 12(1–2):271–290, 1986.

[Mar91] George Edward Martin. *Polyominoes: A Guide to Puzzles and Problems in Tiling*. Mathematical Association of America, 1991.

Printed and bound by CPI Group (UK) Ltd, Croydon, CR0 4YY
23/10/2024
01778227-0005